Heat Transfer Fundamentals for Metal Casting

Second Edition
with SI Units

Heat Transfer Fundamentals for Metal Casting

Second Edition
with SI Units

D.R. Poirier and E.J. Poirier

A Publication of

Minerals • Metals • Materials

A Publication of The Minerals, Metals & Materials Society
420 Commonwealth Drive
Warrendale, Pennsylvania 15086
(412) 776-9000

Library of Congress Catalog Number 94-75872
ISBN Number 0-87339-274-4

If you are interested in purchasing a copy of this book, or if you
would like to receive the latest TMS publications catalog, please
telephone 1-800-759-4867.

FOREWORD

Since 1970 students of metallurgical engineering, materials science and engineering, and related engineering disciplines typically study transport phenomena as a required course in B.S. curricula in the United States. In subsequent courses, however, the concepts learned in transport phenomena may not be reinforced by application because of the many constraints on engineering curricula. Specifically, exposure to quantitative aspects of materials processing, in which transport phenomena plays a strong role, is often overlooked in undergraduate programs of study. Consequently, after students become engineers involved in materials processing and manufacturing, they do not draw on their knowledge of transport phenomena and apply it to process improvements.

The intent of this book is to get process engineers and manufacturing engineers in the metal casting industry to apply quantitative fundamentals of heat transfer. There are many computer simulation software packages now commercially available, and some casting engineers are using these as tools, along with empirical experience, to design castings and casting methods. This book does not fall in this category, but we do hope that it serves as a primer to casting engineers who would like to take that road but do not quite know how to get started.

We have selected topics and examples to illustrate how the physical and transport properties of the casting and mold materials interact to control solidification heat transfer. One can also see the effects of process variables, such as mold and metal temperatures and heat transfer coefficients, on solidification times. An understanding of these variables can assist casting engineers in making economical products of high quality.

To the state-of-the-art engineers and researchers already deeply involved in applying transport phenomena and computer simulations, we simply state this book is not for you. For the sake of competitiveness in manufacturing, we hope your numbers will increase, and maybe this will help. We consider this to be a teaching book; we have interlaced the entire presentation with numerous examples. The book could be used as part of an undergraduate course in materials processing or manufacturing, or it could be used by any engineer with the willingness to do some "self-teaching."

It would be good if the reader has a basic knowledge of heat transfer, but if not then at least he or she should be able to use calculus. The first two chapters review heat transfer and a few concepts pertaining to solidification. After that the reader will see that the emphasis is on conduction heat transfer with a pedagogy in casting solidification. The final two chapters show how heat transfer during solidification can be used in riser design and give an introduction to numerical methods to inspire some confidence in students or casting engineers who want to venture further.

Tucson, Arizona   DRP
Gales Ferry, Connecticut   EJP

December 1993

PREFACE

The first edition made use of English units in the many examples throughout the text. Professor Campbell of the University of Birmingham has taken this choice of units to be a disadvantage. To quote him:

> "It is fervently to be hoped that a second version of the text be produced using SI units. It would then be a pleasure to recommend the book widely. (It would also probably be welcomed by US students.)"

We reconsidered and enthusiastically agree with Professor Campbell. We relayed this matter of units to the publisher (The Minerals, Metals and Materials Society) and requested their support of the second edition with SI units. By the way, Professor Campbell warns the unwary student not to mix water and hot steel.

Tucson, Arizona DRP
Gales Ferry, Connecticut EJP

December 1993

ACKNOWLEDGEMENTS

The authors express their sincere thanks to Dr. Gordon Geiger who alerted us that The Minerals, Metals and Materials Society (TMS) might be willing to publish this book. He went further and urged the Publication Committee of TMS to consider this book. Originally the manuscript for the first edition was prepared for the Investment Casting Institute (ICI) for a short course on gating and risering. We thank ICI for partially funding the preparation of the first edition. The second edition was typed by Mary Cromwell and illustrated by Alison Habel, both at the University of Arizona. The authors appreciate the great job each performed so ably and willingly.

TABLE OF CONTENTS

I. TYPES OF HEAT TRANSFER

Heat is transferred by three basic mechanisms: conduction, convection and radiation. In the context of metal casting, conduction is the mechanism by which heat is transferred internally within the solidifying metal and the mold. Convection within a solidifying metal (i.e., the all-liquid zone) can have some important consequences, e.g., the heating of gates during pouring and macrosegregation in thick castings. In investment castings, convection relates to the rate of heat loss from the outside surface of a shell-mold to the surroundings, and it is in this context that heat transfer by convection is considered, herein. Radiation heat transfer can be important, because heat loss from open risers or from the surfaces of preheated shell molds is largely by radiation.

Each of the three types of heat transfer is described in the following sections.

A. Conduction

There can be steady state conduction or transient conduction. In casting applications transient conduction is by far the more prevalent scenario. However, it is important to understand steady state conduction as a background to understanding transient conduction.

Consider the plane wall shown in Fig. 1. The surface at $x = 0$ is hotter than the surface at $x = L$, so heat is conducted from left-to-right according to

$$Q = \frac{kA}{L} (T_0 - T_L) \tag{1}$$

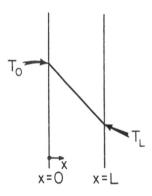

Figure 1: Steady state temperature distribution in a plane wall.

Heat Transfer Fundamentals for Metal Casting
Second Edition with SI Units
Edited by D.R. Poirier and E.J. Poirier
The Minerals, Metals & Materials Society, 1994

where

Q = heat flow rate, W

A = area of the plate, m^2

L = thickness of the plate, m

T_0, T_L = temperatures, K

k = thermal conductivity of the plate, W m^{-1} K^{-1}

The thermal conductivity is a physical property, with a value that depends upon the material and its temperature. Some data are given in Table 1 to illustrate typical values for solids, liquids, and gases.

Example 1

An insulating wall of glass wool is 150 mm thick with an area of 10 m^2. Calculate the rate of heat loss (W) through the wall if its two surfaces are maintained at 283 K and 300 K, respectively.

From Table 1, the thermal conductivity is 0.038 W m^{-1} K^{-1}. Then, Eq. (1) gives

$$Q = \frac{0.038 \text{ W}}{\text{m K}} \left| \frac{10 \text{ m}^2}{0.15 \text{ m}} \right| (300-283) \text{ K}$$

$$Q = 43 \text{ W}$$

This is low value, indeed, which is not surprising because glass wool is a very good insulator.

A better way to write Eq. (1) is

$$q = -k \frac{dT}{dx} \tag{2}$$

where

$q = Q/A$ = heat flux, W m^{-2}

$\frac{dT}{dx}$ = temperature gradient, K m^{-1}

Table 1

Thermal Conductivity of Various Materials at 273 K

Material	k $W\ m^{-1}\ K^{-1}$
Metals:	
Silver (pure)	410
Copper (pure)	385
Aluminum (pure)	202
Nickel (pure)	93
Iron (pure)	73
Carbon steel, 1% C	43
Lead (pure)	35
Chrome-nickel steel (18% Cr, 8% Ni)	16.3
Nonmetallic Solids:	
Quartz, parallel to axis	41.6
Magnesite	4.15
Glass wool	0.038
Liquids:	
Mercury	8.21
Water	0.556
Lubricating oil, SAE 50	0.147
Gases:	
Hydrogen	0.175
Helium	0.141
Air	0.024
Water vapor (saturated)	0.0206
Carbon dioxide	0.0146

From J. P. Holman, *Heat Transfer*, sixth edition, McGraw-Hill, New York, NY, 1986, p. 8.

Equation (2) is a form of *Fourier's rate law*, which is the basic equation for analyzing heat conduction. Equation (2) is written for steady state and for unidirectional heat flow. Therefore, temperature is only a function of x and the gradient can be written as a full derivative. The negative sign is needed so that the heat is conducted from "hot" to "cold" and never *vice-versa*.

Consider Fig. 1 again. The gradient is

$$\frac{dT}{dx} = \frac{T_L - T_0}{L - 0} = -\frac{T_0 - T_L}{L}. \tag{3}$$

Then Eqs. (2) and (3) give

$$q = -k\left[-\frac{T_0 - T_L}{L}\right]$$

or

$$q = \frac{Q}{A} = +k\frac{T_0 - T_L}{L}. \tag{4}$$

Of course, Eqs. (1) and (4) are the same. Furthermore, with $T_0 > T_L$ the flux, q, is positive so that heat is conducted from left-to-right (i.e., from "hot" to "cold").

The temperature distribution is linear only when steady state prevails. However, when the situation is transient the temperature distribution is nonlinear (Fig. 2). Consider a thin slice of the material with a thickness of Δx. Notice that the gradient at x is somewhat greater than the gradient at $x + \Delta x$. Therefore, the heat conducted into the thin slice across the surface at x is greater than the heat conducted out of the slice across the surface at $x + \Delta x$. Therefore, the energy (and temperature) within the slice must increase. All of this can be expressed by simply making a mathematical statement of the *conservation of energy*. This is written as

$$Aq\Big|_x = Aq\Big|_{x+\Delta x} + A\Delta x\rho C_p\frac{\partial T}{\partial t} \tag{5}$$

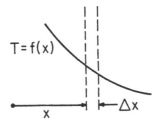

Figure 2: Nonlinear temperature distribution
in transient heat conduction.

4

where

ρ = density, kg m^{-3}

C_p = heat capacity, J kg^{-1} K^{-1}

$\frac{\partial T}{\partial t}$ = rate of temperature increase, K s^{-1}

At this point the two terms with the flux, q, should be familiar. Let's examine the remaining term in more detail. $A\Delta x\rho$ is simply the mass of the slice; therefore the entire third term represents a rate of increase of energy within the slice.

Equation (5) is used to derive the *energy equation* for unidirectional heat conduction. After dividing each term by $A\Delta x$, Eq. (5) is rearranged to the following form:

$$\lim_{\Delta x \to 0} \frac{q\big|_{x+\Delta x} - q\big|_x}{\Delta x} = -\rho C_p \frac{\partial T}{\partial t} \tag{6}$$

The left side of Eq. (6) is the definition of the first derivative; so after the limit is taken, we have

$$\frac{\partial q}{\partial x} = -\rho C_p \frac{\partial T}{\partial t}. \tag{7}$$

Notice that partial derivatives are used, because both q and T are functions of x and t. Similarly, Eq. (2) should be written as

$$q = -k \frac{\partial T}{\partial x} \tag{8}$$

and after combining Eqs. (7) and (8), the *energy equation* for conduction heat transfer results. It is

$$\frac{\partial T}{\partial t} = \frac{1}{\rho C_p} \frac{\partial}{\partial x} \left[k \frac{\partial T}{\partial x} \right] \tag{9}$$

If k is uniform, then

$$\frac{\partial T}{\partial t} = \alpha \frac{\partial^2 T}{\partial x^2} \tag{10}$$

5

where α is the *thermal diffusivity* and defined as

$$\alpha = \frac{k}{\rho c_p}. \qquad (11)$$

The units of α are $m^2 \, s^{-1}$.

Actually Eq. (10) can be used as the starting point for solving either steady state or transient problems. For example, in the steady state plane wall problem of Fig. 1, there is no change of temperature with time so Eq. (10) reduces to

$$\frac{d^2T}{dx^2} = 0. \qquad (12)$$

If the second derivative is zero, then the first derivative must be a constant. Hence

$$\frac{dT}{dx} = C_1$$

and another integration gives

$$T = C_1 x + C_2 \qquad (13)$$

where C_1 and C_2 are integration constants. Notice that the temperature distribution is linear, a result that was previously assumed.

There are two arbitrary constants in Eq. (13) because two integrations were required; they are evaluated by two *boundary conditions*. A boundary condition is a known value of the temperature or the temperature gradient at a particular value of x. For example, suppose we know that

$$T = T_0 \qquad \text{at} \qquad x = 0 \qquad (14)$$

and

$$T = T_L \qquad \text{at} \qquad x = L. \qquad (15)$$

By applying Eqs. (14) and (15) to Eq. (13), we get $C_2 = T_0$ and $C_1 = (T_L - T_0)/L$. Therefore,

$$T = \left[\frac{T_L - T_0}{L} \right] x + T_0. \qquad (16)$$

Immediately we see that the slope or temperature gradient is $(T_L - T_0)/L$ so that the flux (q) and heat flow rate (Q) are

$$q = -k \frac{dT}{dx} = +k \frac{(T_0 - T_L)}{L} \qquad (17)$$

and

$$Q = qA = \frac{kA}{L} (T_0 - T_L) \tag{18}$$

Notice that with Eq. (18), we are back to Eq. (1). A complete circle has been made, but that has not been the intent. Rather by recognizing that Eq. (10) is valid for transient and steady state conduction, it serves as a logical starting point. The other necessary relationship is Eq. (8), which enables us to connect the temperature distribution and the flux.

In *Example 1* you might have wondered how there could be a plane wall of glass wool. Realistically, of course, the glass wool would be between two supporting walls, as depicted in Fig. 3. At steady state, Eq. (12) applies to each of the three layers so that temperature must be linear in each. However, the thermal conductivities are not equal so that the gradients within each are different. With contact thermocouples, we could measure T_1 and T_4, but it is unlikely that we would measure T_2 and T_3. We seek those internal temperatures and the heat loss (Q) through the composite wall.

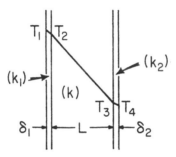

Figure 3: Steady state temperature in a composite wall.

A convenient way to attack the problem is to use Ohm's law as an analog to Eq. (18). Ohm's law is

$$I = \frac{\Delta E}{R} \tag{19}$$

where I is the current (i.e., flow of coulombs), ΔE is the voltage (i.e., potential difference) and R is the resistance. In Eq. (18) Q is flow of heat so it is analogous to I, and $(T_0 - T_L)$ is a temperature drop and analogous to ΔE. That leaves the *thermal resistance*, which is defined as

$$R_T = \frac{L}{kA}. \tag{20}$$

7

Now returning to Fig. 3, there are three thermal resistors: δ_1/k_1A, L/kA and δ_2/k_2A. The three resistors are in series so that the total thermal resistance is

$$R_T = \frac{\delta_1}{k_1A} + \frac{L}{kA} + \frac{\delta_2}{k_2A}. \qquad (21)$$

The temperature drop across the three resistors is $(T_1 - T_4)$ so that the heat flow rate is

$$Q = \frac{(T_1 - T_4)}{R_T}. \qquad (22)$$

Because T_1 is known, we can then determine T_2 because

$$Q = \frac{(T_1 - T_2)}{(\delta_1/k_1A)}, \qquad (23)$$

and similarly T_4 is known so that T_3 can be determined from

$$Q = \frac{(T_3 - T_4)}{(\delta_2/k_2A)}. \qquad (24)$$

Example 2

The 150 mm of glass wool in *Example 1* is supported by 6 mm thick sheets of stainless steel. The external surface temperatures are 300 K and 283 K; determine the rate of heat loss through the composite wall and the internal temperatures.

From Table 1, the thermal conductivity of stainless steel is 16.3 W m^{-1} K^{-1}. The thermal resistance of the wall is

$$R_T = \frac{2 \text{ m K}}{16.3 \text{ W}} \left| \frac{0.006 \text{ m}}{10 \text{ m}^2} \right. + \frac{\text{m K}}{0.038 \text{ W}} \left| \frac{0.15 \text{ m}}{10 \text{ m}^2} \right.$$

$$R_T = (7.36 \times 10^{-5} + 0.395) \text{ K W}^{-1}$$

(Notice that practically all of the resistance is within the glass wool, itself, because it is an excellent insulator.)

Now

$$Q = \frac{(T_1 - T_4)}{R_T} = \frac{(300-283) \text{ K}}{} \left| \frac{\text{W}}{0.395 \text{ K}} \right.$$

$$Q = 43 \text{ W}$$

8

(The result is the same as *Example 1.*)

$$(T_1 - T_2) = \frac{43 \text{ W}}{} \left| \frac{7.36 \times 10^{-5} \text{ K}}{2 \text{ W}} \right. = 0.0016 \text{ K}$$

and
$$(T_3 - T_4) = 0.0016 \text{ K.}$$

(The temperature drop through each sheet of stainless steel is very small because the thermal resistance of each sheet is negligible, when compared to the thermal resistance of the glass wool.)

Now let's consider a transient problem that has application in materials processing, including metal casting. A very thick solid is initially at a uniform temperature, T_i, and then its surface temperature is abruptly raised to and maintained at T_0 (see Fig. 4a). Immediately a very steep temperature gradient is set up at the surface so there must be heat conduction into the solid. Temperature varies with distance x and time t as shown schematically in Fig. 4b.

Temperature must satisfy Eq. (10), which contains a first derivative with respect to time (t) and a second derivative with respect to distance (x). Therefore, to get a particular solution for $T(x,t)$ without any integration constants, we specify an *initial condition* and two boundary conditions. They are as follows:

Initial Condition

$$T(x,0) = T_i, \qquad x \geq 0 \tag{25}$$

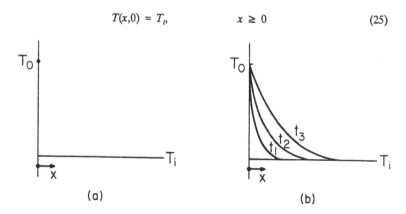

(a) (b)

Figure 4:   Transient temperature in a semi-infinite thick solid: (a) solid with uniform temperature T_i, (b) temperature distributions for times $t_1 < t_2 < t_3$.

9

Boundary Conditions

$$T(0,t) = T_0, \qquad t > 0 \qquad (26)$$

$$T(\infty,t) = T_i, \qquad t \geq 0 \qquad (27)$$

Equation (25) expresses in mathematical form that, at $t = 0$, the temperature is uniform and equal to T_i. Equation (26) states that, at the surface of the solid ($x = 0$), the temperature is maintained at a constant temperature T_0. Equation (27) simply states that the temperature must decrease toward T_i with an increase in x. For mathematical convenience, $x \to \infty$ so the solid is considered to be *semi-infinite*.

The solution that satisfies Eq. (10) and Eqs. (25)-(27) is

$$\frac{T - T_0}{T_i - T_0} = \text{erf}\left[\frac{x}{2\sqrt{\alpha t}} \right] \qquad (28)$$

where the error function is defined as

$$\text{erf}\left[\frac{x}{2\sqrt{\alpha t}} \right] = \frac{2}{\sqrt{\pi}} \int_0^{x/2\sqrt{\alpha t}} e^{-\eta^2}\, d\eta. \qquad (29)$$

Readers with experience in applied statistics might be familiar with the error function. For the uninitiated, simply think of it as you think of any other function (e.g., sine, logarithm, square root, etc.). In other words if the argument of the function is known then the function can be determined from a tabulation.

The error function is given in Table 2. To satisfy ourselves that sensible results are obtained, let's use Eq. (28) and calculate the temperature at $t = 0$. Then the argument of the error function is ∞, and Table 2 indicates that the error function must be 1, its maximum value. Then Eq. (28) gives

$$\frac{T - T_0}{T_i - T_0} = 1,$$

so that $T = T_i$. Of course, we get back the initial condition. Now calculate the temperature at the surface for $t > 0$. The argument of the error function is 0, Table 2 gives an error function of 0, and Eq. (28) gives

$$\frac{T - T_0}{T_i - T_0} = 0.$$

Therefore, $T = T_0$ at $x = 0$, which is one of the boundary conditions. Finally at $x = \infty$, it is easy to see that $T = T_i$, which is the other boundary condition.

Table 2

The Error Function

N	erf N	N	erf N	N	erf N
0.00	0.00000	0.50	0.5205	1.1	0.8802
0.05	0.05637	0.55	0.5633	1.2	0.9103
0.10	0.1125	0.60	0.6039	1.3	0.9340
0.15	0.1680	0.65	0.6420	1.4	0.9523
0.20	0.2227	0.70	0.6778	1.5	0.9661
0.25	0.2763	0.75	0.7112	1.6	0.9763
0.30	0.3286	0.80	0.7421	1.7	0.9838
0.35	0.3794	0.85	0.7707	1.8	0.9891
0.40	0.4284	0.90	0.7969	1.9	0.9928
0.45	0.4755	0.95	0.8209	2.0	0.9953
		1.00	0.8427	∞	1

For casting applications, it is more important to know the temperature gradient than it is to know the temperature distribution given by Eq. (28). Without an exposition of the calculus involved, the temperature gradient obtainable from Eq. (28) is

$$\frac{\partial T}{\partial x} = \frac{T_i - T_0}{\sqrt{\pi \alpha t}} \exp \left[-\frac{x^2}{4\alpha t} \right]. \tag{30}$$

Notice that the gradient varies with both x and t. At a fixed time (e.g., $t = t_2$ in Fig. 4b), the gradient decreases with increasing x; at a particular location, the gradient decreases with increasing time.

Example 3

The surface of a thick mold of plaster, initially at 300 K, is abruptly heated to 800 K and maintained at that temperature. If the surface has an area of 0.1 m², determine the heat flow rate into the mold at 1, 5 and 15 minutes. The properties of the plaster are $k = 0.50$ W m⁻¹ K⁻¹, $\rho = 1440$ kg m⁻³, $C_p = 840$ J kg⁻¹ K⁻¹, and $\alpha = 4.13 \times 10^{-7}$ m² s⁻¹.

All of the heat absorbed by the mold must pass through its surface, so we should evaluate the flux at $x = 0$. This is

$$q_0 = -k \left[\frac{\partial T}{\partial x} \right]_{x=0},$$

11

and with Eq. (30) we get

$$q_0 = \frac{k(T_0 - T_i)}{\sqrt{\pi \alpha t}}. \tag{31}$$

Notice that q_0 varies with time. Then, at $t = 1$ min.,

$$Q_0 = q_0 A = \frac{kA(T_0 - T_i)}{\sqrt{\pi \alpha t}}$$

$$Q_0 = \frac{(0.50)(0.1)(800-300)}{\left(\pi \times 4.13 \times 10^{-7} \times 60\right)^{1/2}} = 2833 \text{ W}$$

Similar calculations are done for, $t = 5$, and 15 min. The results are plotted in Fig. 5.

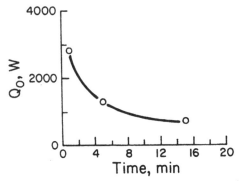

Figure 5: Rate of heat conducted into a plaster mold.

Example 4

Estimate the thickness of plaster mold in *Example 3* so that it can be considered to be semi-infinite thick.

The heated surface of the mold is at $x = 0$. Let the thickness of the mold be L; then when the temperature at $x = L$ increases, the mold can no longer be considered semi-infinite thick. Reference to Table 2 shows that the error function is practically at its maximum value of 1 when $N = 2$. Therefore,

$$\frac{x}{2\sqrt{\alpha t}} = 2.$$

We set $x = L$ and solve for L, when $t = 60$ s.

$$L = (2)(2)(4.13 \times 10^{-7} \times 60)^{1/2}$$

$$L = 0.020 \text{ m} = 20 \text{ mm}$$

For $t = 5$ min and 15 min, the results are $L = 44.5$ mm and 77.1 mm, respectively. If the mold has a thickness less than these calculated values of L, then it is not semi-infinite, and the solution for the error function is invalid. With the argument selected as 2, then at $x = L$

$$\frac{T - T_0}{T_i - T_0} = \frac{T - 800}{300 - 800} = 0.995322$$

and $T = 302.3$ K, which is sufficiently close to the initial temperature that the mold is semi-infinite thick. If we had selected a more stringent requirement (e.g., $x/2\sqrt{\alpha t} = 3$), then the calculated values of L would have been smaller.

B. Convection

It is well known that a heated solid will cool faster when placed in front of a fan than when it is placed in still air. Cooling is said to be by *convection heat transfer*. To deal with convection quantitatively, we must specify or know the velocity, type of convection, the thermal properties of the fluid, and the geometry of the solid surface. For now we qualitatively describe convection heat transfer and relate it to conduction.

Consider the heated plate shown in Fig. 6, which is being cooled by the passing fluid. The velocity of the fluid is shown as $v_x(y)$; notice that the velocity is zero at the surface of the plate because of the viscous nature of the fluid. Exactly at the surface, therefore, heat is transferred only by conduction, and the flux at the surface (q_0) is

$$q_0 = -k \left[\frac{\partial T}{\partial x} \right]_{x=0}. \tag{32}$$

where the gradient is in the fluid. If we are able to calculate the gradient at the surface, then we can get the heat transfer from the plate to the cooling fluid and, in turn, calculate the temperature distribution within the plate. However, that is a "big if," because the gradient depends upon the details of the fluid flow, which can be analyzed but only with great effort. Therefore, we almost always rely on empirical data or on results of previous analyses of the fluid flow to get the information required to solve the heat transfer.

To express the effect of convection we use

$$q_0 = h \left(T_0 - T_\infty \right) \tag{33}$$

13

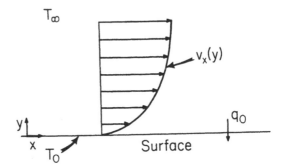

Figure 6: Convection heat transfer to a surface.

where

T_0 = surface temperature,

and

T_∞ = temperature of the fluid away from the surface.

The quantity h is called the *convection heat transfer coefficient* or simply heat transfer coefficient. The units of h are W m^{-2} K^{-1}. Typical values are given in Table 3.

In Table 3, reference is made to *forced* convection and *free* convection. A fan blowing air, water flowing in a pipe, and relative motion between a solid in a fluid are examples of forced convection. If a heated plate is exposed to ambient air, there is a movement of air as a result of density gradients near the plate. This type of convection is called free or natural convection.

Example 5

Air at 300 K blows past a heated face of a slab with dimensions of 150 mm by 150 mm. The slab is maintained at 425 K and $h = 55$ W m^{-2} K^{-1}. Calculate the heat transfer rate.

Equation (33) is used as follows:

$$Q = Aq_0 = hA\left(T_0 - T_\infty\right)$$

$$= \frac{55 \text{ W}}{\text{m}^2 \text{ K}} \left| 2.25 \times 10^{-2} \text{ m}^2 \right| (425-300)\text{K} \qquad (34)$$

$$= 154.7 \text{ W}$$

14

Table 3

Convection Heat Transfer Coefficients

Mode	h
	W m^2 K^{-1}
Free convection, $\Delta T = 30$ K	
Vertical plate, 0.3 m high, in air	4.5
Horizontal cylinder, 50 mm diameter, in air	6.5
Horizontal cylinder, 20 mm diameter, in water	890
Forced convection	
Airflow at 2 m s^{-1} over 0.2 m square plate	12
Airflow at 35 m s^{-1} over 0.75 m square plate	75
	65
Air at 2 atm flowing in 25 mm diameter tube at 10 m s^{-1}	3500
Water at 0.5 kg s^{-1} flowing in 25 mm diameter tube	180
Airflow across 50 mm diameter cylinder with velocity of 50 m s^{-1}	
Boiling water	
In a pool or container	2500-35 000
Flowing in a tube	5000-100 000

From J. P. Holman, *Heat Transfer*, sixth edition, McGraw-Hill, New York, NY, 1986, p. 13.

Figure 7 depicts an interaction between convection and conduction. In Fig. 1 the surface temperatures were known *a priori* and so they could be specified. More often than not, however, the temperatures of the fluids ($T_{\infty 1}$ and $T_{\infty 2}$) are known, and then we seek the temperature distribution within the solid that depends upon the temperatures $T_{\infty 1}$ and $T_{\infty 2}$ and the corresponding heat transfer coefficients.

At steady state, the temperature distribution within the plate is linear so that

$$h_1 \left(T_{\infty 1} - T_0 \right) = \frac{k}{L} \left(T_0 - T_L \right) \tag{35}$$

and

$$h_2 \left(T_L - T_{\infty 2} \right) = \frac{k}{L} \left(T_0 - T_L \right). \tag{36}$$

15

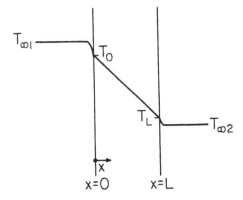

Figure 7: Steady state temperature distribution in a plane wall
between two fluids at different temperatures.

The two unknown temperatures, T_0 and T_L, can be determined by simultaneously solving Eqs. (35) and (36).

An easier solution method is to use the electrical analog for Eq. (33); i.e.,

$$Q = \frac{(T_0 - T_\infty)}{R_T}$$

where R_T is the thermal resistance for convection heat transfer and is defined as

$$R_T = \frac{1}{hA}. \tag{37}$$

In Fig. 7, there are three resistors in series, so the total resistance is

$$R_T = \frac{1}{h_1 A} + \frac{L}{kA} + \frac{1}{h_2 A}. \tag{38}$$

The temperature difference is $T_{\infty 1} - T_{\infty 2}$, so that

$$Q = \frac{T_{\infty 1} - T_{\infty 2}}{R_T}, \tag{39}$$

and the unknown temperatures are

$$\frac{T_{\infty 1} - T_0}{T_{\infty 1} - T_{\infty 2}} = \frac{1}{h_1 A R_T} = \left[1 + \frac{h_1 L}{k} + \frac{h_1}{h_2} \right]^{-1} \tag{40}$$

16

and

$$\frac{T_L - T_{\infty 2}}{T_{\infty 1} - T_{\infty 2}} = \frac{1}{h_2 A R_T} = \left[1 + \frac{h_2 L}{k} + \frac{h_2}{h_1} \right]^{-1} \qquad (41)$$

Example 6

A 6 mm thick plate of stainless steel ($k = 16.3$ W m^{-1} K^{-1}) is at steady state with moving air on one side ($h_1 = 30$ W m^{-2} K^{-1}, $T = 300$ K) and stagnant air on the other ($h_2 = 6$ W m^{-2} K^{-1}, $T = 280$ K). Calculate the heat flux and the surface temperatures.

The thermal resistance is given by Eq. (38).

$$R_T = \left[\frac{1}{30} + \frac{6 \times 10^{-3}}{16.3} + \frac{1}{6} \right] \frac{1}{A}$$

$$= \frac{0.2004}{A} \text{ K W}^{-1}$$

Then the flux is

$$q = \frac{Q}{A} = \frac{T_{\infty 1} - T_{\infty 2}}{A R_T}$$

$$= \frac{(300-280) \text{ K W}}{0.2004 \text{ K}} = 99.80 \text{ W}$$

Equations (40) and (41) can be used to calculate the surface temperatures.

$$T_{\infty 1} - T_0 = \left[1 + \frac{(30)(6 \times 10^{-3})}{16.3} + \frac{30}{6} \right]^{-1} (300-280) = 3.33 \text{ K}$$

$$T_L - T_{\infty 2} = \left[1 + \frac{(6)(6 \times 10^{-3})}{16.3} + \frac{6}{30} \right]^{-1} (300-280) = 16.64 \text{ K}$$

Therefore, the surface temperatures are 296.67 K and 296.64 K.

The field of heat transfer cuts across many disciplines and, consequently, over the years much effort has been devoted to experiments and analyses on convection heat transfer. A few of the many results are presented here.

For laminar flow parallel to the surface of a plate, the heat transfer coefficient is calculated from

$$Nu_L = 0.664 \ Re_L^{1/2} \ Pr^{1/3}. \tag{42}$$

Equation (42) is typical for forced convection, in that the heat transfer coefficient is correlated in terms of three dimensionless groups: a *Nusselt* number (Nu_L), a *Reynolds* number (Re_L), and a *Prandtl* number (Pr). They are defined as follows:

$$Nu_L = \frac{hL}{k}; \tag{43}$$

$$Re_L = \frac{LV_\infty \rho}{\mu}; \tag{44}$$

and

$$Pr = \frac{C_p \mu}{k}, \tag{45}$$

where L is the length of the plate, V_∞ is the free-stream velocity of the fluid, ρ is density, μ is viscosity, C_p is heat capacity, k is thermal conductivity, and all thermal properties (k, ρ, μ, and C_p) are for the fluid. Inspection of the dimensionless groups shows that Eq. (42) embodies the effect of a particular fluid, the length of the plate, and the velocity of the fluid on the heat transfer coefficient.

The properties of some fluids encountered in metal casting are given in Tables 4 through 6. Notice that the fluid properties vary somewhat with temperature; this is especially so for gases. When there is an appreciable difference between the surface temperature of the solid (T_0) and the free-stream temperature (T_∞) of the fluid, it is recommended that the properties be evaluated at the so-called *film temperature*. The film temperature is defined as

$$T_f = \frac{T_0 + T_\infty}{2}. \tag{46}$$

When using correlations, such as Eq. (42), one must be aware of its limitations. Equation (42) is valid provided $0.6 \leq Pr \leq 50$. It does not apply to fluids with very low Prandtl numbers, like liquid metals, or to fluids with very high Prandtl numbers, like heavy oils. Equation (42) is also restricted to laminar flows so that $Re_L < 10^6$ (approx.). If flow is turbulent, then

18

$$Nu_L = \frac{1}{2} Re_L Pr^{1/3} \left[\frac{0.455}{(\log Re_L)^{2.584}} - \frac{3300}{Re_L} \right] \tag{47}$$

for $Re_L > 10^7$.

Table 4

Thermal Properties of Air at Atmospheric Pressure

T, K	ρ, kg m^{-3}	C_p, kJ kg^{-1} K^{-1}	μ, kg m^{-1} s^{-1} $\times 10^5$	k, W m^{-1} K^{-1}	α, m^2 s^{-1} $\times 10^4$	Pr
300	1.1774	1.0057	1.8462	0.02624	0.22160	0.708
400	0.8826	1.0140	2.286	0.03365	0.3760	0.689
500	0.7048	1.0295	2.671	0.04038	0.5564	0.680
600	0.5879	1.0551	3.018	0.04659	0.7512	0.680
700	0.5030	1.0752	3.332	0.05230	0.9672	0.684
800	0.4405	1.0978	3.625	0.05779	1.1951	0.689
900	0.3925	1.1212	3.899	0.06279	1.4271	0.696
1000	0.3524	1.1417	4.152	0.06752	1.6779	0.702
1200	0.2947	1.179	4.69	0.0782	2.251	0.707
1400	0.2515	1.214	5.17	0.0891	2.920	0.705
1600	0.2211	1.248	5.63	0.100	3.609	0.705
1800	0.1970	1.287	6.07	0.111	4.379	0.704
2000	0.1762	1.338	6.50	0.124	5.260	0.702
2200	0.1602	1.419	6.93	0.139	6.120	0.707

From F. P. Incropera and D. P. DeWitt, *Fundamentals of Heat and Mass Transfer*, third edition, John Wiley & Sons, New York, NY, 1990, p. A15.

The values of μ, k, C_p, and Pr are not strongly pressure-dependent and may be used up to pressures of 10×10^5 N m^{-2}.

Heat transfer coefficients for free convection are represented in the following form for many circumstances:

$$Nu = C(Gr Pr)^m \tag{48}$$

where C and m are constants that are given in Table 7.

In general, the Nusselt number contains a characteristic length of the solid surface. For free convection next to vertical planes and vertical cylinders, the characteristic length is the vertical length, L. For horizontal cylinders, the diameter, D, is the characteristic length.

Table 5

Thermal Properties of Water

T, K	C_p, kJ kg⁻¹ K⁻¹	ρ, kg m⁻³	μ, kg m⁻¹ s⁻¹	k, W m⁻¹ K⁻¹	Pr	$\dfrac{g\beta\rho^2 C_p}{\mu k}$, m⁻³ K⁻¹
273	4.217	1000	1.75×10^{-3}	0.569	12.99	-2.83×10^{6}
280	4.198	1000	1.42×10^{-3}	0.582	10.26	2.29×10^{6}
290	4.184	999	1.08×10^{-3}	0.598	7.56	1.10×10^{7}
300	4.179	997	8.55×10^{-4}	0.613	5.83	2.15×10^{7}
310	4.178	993	6.95×10^{-4}	0.628	4.62	3.35×10^{7}
320	4.180	989.1	5.77×10^{-4}	0.640	3.77	4.74×10^{7}
330	4.184	984.2	4.89×10^{-4}	0.650	3.15	6.30×10^{7}
340	4.188	979.4	4.20×10^{-4}	0.660	2.66	8.04×10^{7}
350	4.195	973.7	3.65×10^{-4}	0.668	2.29	9.99×10^{7}
360	4.203	967.1	3.24×10^{-4}	0.674	2.02	1.23×10^{8}
370	4.214	963.4	2.89×10^{-4}	0.679	1.80	1.42×10^{8}

From F. P. Incropera and D. P. DeWitt, *Fundamentals of Heat and Mass Transfer*, third edition, John Wiley & Sons, New York, NY, 1990, p. A22.

Note: $Gr_x Pr = \left[\dfrac{g\beta\rho^2 C_p}{\mu k} \right] x^3 \, \Delta T$

Table 6

Thermal Properties of Low Melting Point Metals

Metal	Melting Point, K	Temperature, K	ρ, kg m^{-3} $\times 10^{-3}$	μ, kg m^{-1} s^{-1} $\times 10^4$	C_p, kJ kg^{-1} K^{-1}	k, W m^{-1} K^{-1}	Pr
Bismuth	544	589	10.01	1.62	0.144	16.4	0.014
		1033	9.47	0.79	0.165	15.6	0.0084
Lead	600	644	10.5	2.40	0.159	16.1	0.024
		977	10.1	1.37	0.155	14.9	0.016
Lithium	452	477	0.51	0.60	4.19	38.1	0.065
		1255	0.44	0.42	4.19		
Mercury	234	283	13.6	1.59	0.138	8.1	0.027
		589	12.8	0.86	0.134	14.0	0.0084
Potassium	336	422	0.81	0.37	0.796	45.0	0.0066
		977	0.67	0.14	0.754	33.1	0.0031
Sodium	370	477	0.90	0.43	1.34	80.3	0.0072
		977	0.78	0.18	1.26	59.7	0.0038
Sodium Potassium: 22% Na	292	366	0.848	0.49	0.946	24.4	0.019
		1033	0.69	0.146	0.883		
56% Na	262	366	0.89	0.58	1.13	25.6	0.026
		1033	0.74	0.16	1.04	28.9	0.058
Lead Bismuth: 44.5% Pb	398	561	10.3	1.76	0.147	10.7	0.024
		922	9.84	1.15			

From F. P. Incropera and D. P. DeWitt, *Fundamentals of Heat and Mass Transfer*, third edition, John Wiley & Sons, New York, NY, 1990, p. A24.

The *Grashof* number, Gr, in Eq. (48) is a measure of the strength of the free convection, whereas the Reynolds number, Re_L, is a measure of the strength of forced convection. The Grashof number is

$$Gr = \frac{g\beta\rho^2(T_0 - T_\infty)L^3}{\mu^2} \qquad (49)$$

Table 7

Constants for Use with Eq. (48) in Free Convection Situations

Geometry	$GrPr$	C	m
Vertical planes and cylinders	10^4-10^9	0.59	1/4
	10^9-10^{13}	0.021	2/5
	10^9-10^{13}	0.10	1/3
Horizontal cylinders	0-10^{-5}	0.4	0
	10^4-10^9	0.53	1/4
	10^9-10^{12}	0.13	1/3
	10^{-10}-10^{-2}	0.675	0.058
	10^{-2}-10^2	1.02	0.148
	10^2-10^4	0.850	0.188
	10^4-10^7	0.480	1/4
	10^7-10^{12}	0.125	1/3
Upper surface of heated plates or lower surface of cooled plates*	2×10^4-8×10^6	0.54	1/4
Upper surface of heated plates or lower surface of cooled plates*	8×10^6-10^{11}	0.15	1/3
Lower surface of heated plates or upper surface of cooled plates*	10^5-10^{11}	0.27	1/4
Vertical cylinder, height = diameter characteristic length = diameter	10^4-10^6	0.775	0.21

From J. P. Holman, *Heat Transfer*, sixth edition, McGraw-Hill, New York, NY, 1986, p. 333.

* $L = \sqrt{\ell_1\ell_2}$, for rectangular plates with dimensions ℓ_1 by ℓ_2.

where β is the volume coefficient of expansion and g is gravitational acceleration. For ideal gases

$$\beta = \frac{1}{T} \qquad (50)$$

where T is the absolute temperature. Values of $GrPr$ for water are conveniently obtainable from the last column in Table 5.

Example 7

A vertical plate (0.3 m by 0.3 m) is maintained at 433 K and is exposed to air at 300 K. Compare the heat transfer coefficients for free convection and forced convection with a velocity of 6 m s⁻¹.

First we gather the properties of air at T_f = ½(433 + 300) = 366 K. From interpolations in Table 4:

ρ = 0.9828 kg m⁻³

C_p = 1.011 kJ kg⁻¹ K⁻¹

μ = 2.136 × 10⁻⁵ kg m⁻¹ s⁻¹

k = 0.03113 W m⁻¹ K⁻¹

Pr = 0.695

Also

β = $1/T_f$ = 1/366 = 2.73 × 10⁻³ K⁻¹

g = 9.81 m s⁻²

Then

$$Gr = \frac{g\beta\rho^2 (T_0 - T_\infty) L^3}{\mu^2}$$

$$= \frac{9.81 \text{ m}}{\text{s}^2} \left| \frac{2.73 \times 10^{-3}}{\text{K}} \right| \frac{0.9828^2 \text{ kg}^2}{\text{m}^6} \left| (433 - 300) \text{ K} \right.$$

$$\frac{0.3^3 \text{ m}^3}{} \left| \frac{\text{m}^2 \text{ s}^2}{(2.136 \times 10^{-5})^2 \text{ kg}^2} \right. = 2.036 \times 10^8$$

$$Re_L = \frac{LV_\infty \rho}{\mu}$$

$$= \frac{0.3 \text{ m}}{} \left| \frac{6 \text{ m}}{\text{s}} \right| \frac{0.9828 \text{ kg}}{\text{m}^3} \left| \frac{\text{m s}}{2.136 \times 10^{-5} \text{ kg}} \right.$$

$$= 8.282 \times 10^4$$

For *free convection*, $Gr \cdot Pr = 1.42 \times 10^8$ and Table 7 gives $C = 0.59$ and $m = 1/4$. Then, using Eq. (48):

$$Nu_L = \frac{hL}{k} = 0.59 \left(1.42 \times 10^8\right)^{1/4}$$

$$= 64.35$$

and

$$h = \frac{(64.35)(0.03113)}{0.3} = 6.68 \text{ W m}^{-2} \text{ K}^{-1}$$

For *forced convection*, $Re_L < 10^6$ so that Eq. (42) applies. Then

$$Nu_L = \frac{hL}{k} = 0.664 \left(8.282 \times 10^4\right)^{1/2} (0.695)^{1/3}$$

$$= 169.3$$

$$\text{Therefore,} \quad h = \frac{(169.3)(0.03113)}{0.3} = 17.6 \text{ W m}^{-2} \text{ K}^{-1}$$

Notice that the heat transfer coefficient for forced convection is almost three times that for free convection.

C. *Radiation*

In contrast to conduction and convection, which involve transport of energy through a material, energy may also be transferred through a vacuum. The mechanism is by electromagnetic radiation, and is specifically called *thermal radiation*.

An ideal thermal radiator, or *blackbody*, emits energy at a rate proportional to the fourth power of the absolute temperature of the body and directly proportional to its surface area. Thus

$$Q = \sigma A T^4 \tag{51}$$

where σ is called the Stefan-Boltzmann constant with the value of 5.670×10^{-8} W m^{-2} K^{-4}. Equation (51) is called the Stefan-Boltzmann law of thermal radiation, and it applies only to blackbodies.

The net radiant *exchange* between two surfaces, expressed as a flux, is proportional to the difference in absolute temperatures to the fourth power; i.e.,

$$q \propto \sigma \left(T_1^4 - T_2^4\right) \tag{52}$$

We have mentioned that a blackbody is a body which radiates energy according to the T^4 law. We call such a body *black* because black surfaces, like a piece of metal covered with carbon black, approximate this type of behavior. Real surfaces, like a polished metal plate, do not radiate as much energy as the blackbody and are approximated as *gray* surfaces. However, the total radiation emitted by these bodies still generally follows the T^4 proportionality. To take account of the gray nature of such surfaces we introduce another factor into Eq. (51), called the emissivity ϵ, which is the ratio of the radiation of the gray surface to that of an ideal black surface. In addition, we must take into account the fact that not all the radiation leaving one surface reaches the other surface because electromagnetic radiation travels in straight lines and some will be lost to the surroundings. We therefore introduce two new factors in Eq. (52) to take into account both situations, so that

$$Q = F\sigma A \left(T_1^4 - T_2^4\right) \qquad (53)$$

where F is a function of the emissivities and the geometries of the two surfaces. The determination of the form of this function for specific configurations is a major subject of radiation heat transfer. Here, we consider only a few cases that can be applied to metal casting. In general, however, radiation heat transfer can be very complex.

A simple radiation problem is encountered when we have a surface, at temperature T_1, completely enclosed by a much larger surface maintained at T_2. The net radiant exchange in this case can be calculated with

$$Q = \epsilon_1 \sigma A_1 \left(T_1^4 - T_2^4\right) \qquad (54)$$

Another situation is depicted in Fig. 8, where there are two radiating circular discs that are separated by a space bounded with an insulator. The net radiant exchange is given by Eq. (53) where A = area of each disc ($\pi d^2/4$) and

$$F = \left[\frac{2 - 2F_{12}}{1 - F_{12}^2} + \frac{1}{\epsilon_1} + \frac{1}{\epsilon_2} - 2 \right]^{-1} \qquad (55)$$

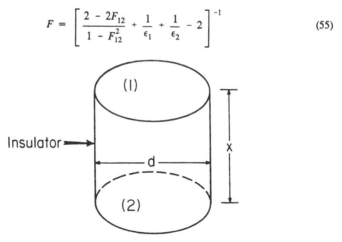

Figure 8: Parallel discs separated by an insulated space.

In Eq. (55), F_{12} is called the *view factor*; it is given in Fig. 9. Emissivities of various materials are in Table 8.

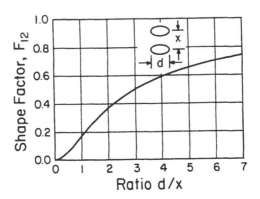

Figure 9: Shape factor for radiation between parallel circular disks. (From J. P. Holman, *Heat Transfer*, sixth edition, McGraw-Hill, New York, NY, 1986, p. 388.)

Example 8

A shell mold with a surface area of 0.1 m² is preheated to 925 K. When it is removed from the preheat furnace, what is the rate of heat loss? Assume that the emissivity of the mold is 0.5.

In this situation, Eq. (54) applies with $\epsilon_1 = 0.5$, $A_1 = 0.1$ m², $T_1 = 925$ K, and $T_2 = 300$ K. Then

$$Q = (0.5)(5.670)(0.1) \left[\left(\frac{925}{100} \right)^4 - \left(\frac{300}{100} \right)^4 \right]$$

$$= 2.05 \times 10^3 \text{ W}$$

Notice that the value of σ (5.670 × 10⁻⁸ W m⁻² K⁴) has been split into two parts: 5.670 and each temperature divided by 100. At this relatively high temperature, the radiation heat transfer is approximately 2 to 4 times that for convection heat transfer. As a general rule, when $T \geq 800$ K (approx.) radiation heat transfer usually dominates.

Table 8

Emissivities of Various Surfaces

Surface	K	Emissivity, ϵ
Aluminum:		
Highly polished plate, 98.3% pure	500-850	0.039-0.057
Commercial sheet	373	0.09
Heavily oxidized	421-778	0.20-0.31
Brass:		
Highly polished:		
73.2% Cu, 26.7% Zn	520-630	0.028-0.031
62.4% Cu, 36.8% Zn, 0.4% Pb, 0.3% Al	530-650	0.033-0.037
82.9% Cu, 17.0% Zn	550	0.030
Hard-rolled, polished, but direction of polishing visible	294	0.038
Dull plate	322-622	0.22
Iron and steel (not including stainless):		
Steel, polished	373	0.066
Iron polished,	700-1300	0.14-0.38
Cast iron, newly turned	295	0.44
turned and heated	1155-1261	0.60-0.70
Mild steel	505-1339	0.20-0.32
Sheet steel with strong, rough oxide layer	297	0.80
Magnesium, magnesium oxide	550-1100	0.55-0.20
Monel metal, oxidized at 1110°F	472-872	0.41-0.46
Nickel:		
Polished	373	0.072
Nickel oxide	922-1528	0.59-0.86
Nickel alloys:		
Copper nickel, polished	373	0.059
Nichrome wire, bright	322-1272	0.65-0.79
Nichrome wire, oxidized	322-772	0.95-0.98
Stainless steels:		
Polished	373	0.074
Type 301; B	505-1214	0.54-0.63
Alumina (85-99.5%, Al_2O_3, 0-12% SiO_2, 0-1% Ge_2O_3); effect of mean grain size, microns (μm):		
10 μm		0.30-0.18
50 μm		0.39-0.28
100 μm		0.50-0.40
Asbestos, board	296	0.96
Brick:		
Red, rough, but no gross irregularities	294	0.93
Fireclay	1273	0.75
Quartz, rough, fused	294	0.93

Selected from J. P. Holman, *Heat Transfer*, sixth edition, McGraw-Hill, New York, NY, 1986, pp. 648-649.

In the development of convection heat transfer, we found it convenient to define a heat transfer coefficient by

$$Q_{conv} = hA_1(T_0 - T_\infty)$$

Because radiation is often closely associated with convection, and the total heat transfer by both convection and radiation is sought, it is worthwhile to put both processes on a common basis by defining a radiation heat-transfer coefficient h_r as

$$Q_{rad} = h_r A_1(T_1 - T_2) \tag{56}$$

where T_1 and T_2 are the absolute temperatures of the two bodies exchanging heat by radiation. The total heat transfer is then the sum of the convection and radiation,

$$Q = (h + h_r)A_1(T_0 - T_\infty) \tag{57}$$

if we assume that the second-radiation-exchange surface is at the same temperature as the fluid (i.e., $T_2 = T_\infty$ and $T_1 = T_0$). For example, the heat loss by free convection and radiation from a hot steam pipe passing through a room could be calculated from Eq. (57).

In many instances the convection heat transfer coefficient is not strongly dependent on temperature. However, this is not so with the radiation heat-transfer coefficient. The value of h_r, corresponding to Eq. (53) is

$$h_r = F\sigma(T_1^2 + T_2^2)(T_1 + T_2) \tag{58}$$

Obviously, the radiation coefficient is a very strong function of temperature.

Example 9

Estimate the rate of heat loss from the top of an open riser (50 mm dia.) of molten steel. The freezing point of the steel is 1773 K, and the emissivity of molten steel is 0.28.

The total heat loss is by free convection and by radiation. Consider convection first and gather the properties for air at $T_f = 1036$ K. From Table 4:

$\rho = 0.3420$ kg m^{-3}

$\mu = 4.249 \times 10^{-5}$ kg m^{-1} s^{-1}

$k = 0.06944$ W m^{-1} K^{-1}

$Pr = 0.703$

28

Also

$$\beta = 1/T_f = 1/1036 = 9.65 \times 10^{-4} \text{ K}^{-1}$$

$$L \simeq \text{diameter} = 0.050 \text{ m}$$

$$g = 9.81 \text{ m s}^{-2}$$

Then

$$Gr = \frac{(9.81\ )(9.65 \times 10^{-4})(0.3420)^2 (1773-300)(0.050)^3}{(4.249 \times 10^{-5})^2}$$

$$= 1.129 \times 10^5$$

$$Gr \cdot Pr = 7.94 \times 10^4$$

Table 7 gives $C = 0.54$ and $m = 1/4$.

$$Nu = \frac{hL}{k} = 0.54(7.94 \times 10^4)^{1/4} = 9.06$$

and

$$h = \frac{(9.06)(0.06944)}{0.05} = 12.6 \text{ W } m^{-2} \text{ K}^{-1}$$

For the radiation heat transfer coefficient, we use Eq. (58) with $F = \epsilon = 0.28$, $T_1 = 1773$ K and $T_2 = 300$ K. Then,

$$h_r = (0.28)(5.670 \times 10^{-8})(1773^2 + 300^2)(1773 + 300)$$

$$= 106.4 \text{ W } m^{-2} \text{ K}^{-1}$$

The total heat loss is

$$Q = (h + h_r) A (T_1 - T_2)$$

$$= (12.6 + 106.4) \left[\frac{\pi}{4} \times 0.05^2 \right] (1773-300)$$

$$= 344 \text{ W}$$

Notice that 89.4% of the heat loss from the top of the riser is by radiation. With a metal with a lower melting point (e.g., Al at 933 K), the heat loss is only 41.3 W and the percentage by radiation is reduced to 63.4%.

II. HEAT TRANSFER IN A SOLIDIFYING METAL

A. *Solidification of Pure Metal*

To determine the time for a casting to solidify, one can estimate the energy absorbed by the mold and equate that energy to the energy extracted from the solidifying metal. Thus there must be an accounting of the energy in the metal. In this section we consider a pure metal.

The energy of the metal is expressed in terms of its *enthalpy*. The enthalpy of a substance has no absolute value, so we use its enthalpy relative to a reference state. As an example, consider iron. Like many elements, iron is allotropic. Below 1184 K, its crystalline form is body centered cubic (b.c.c.). Upon heating at 1184 K it transforms to face centered cubic (f.c.c.), and with continued heating to 1665 K, it transforms back to b.c.c. Finally, it melts at 1809 K.

With the reference state taken as b.c.c. at 298 K, the enthalpy is 0. Then the enthalpy of iron can be represented as shown in Fig. 10. Notice at each transformation temperature, there is a jump in the enthalpy. For example, consider the transformation at 1184 K; the reaction is

$$\alpha \text{ (b.c.c.)} = \gamma \text{ (f.c.c.)}$$

and upon heating through 1184 K, the change in the enthalpy is only

$$\Delta H = H_\gamma - H_\alpha = 941 \text{ J mol}^{-1}$$

Now consider the solidification reaction at 1809 K. Upon cooling liquid transforms to solid, labeled δ, and the associated change in enthalpy is

$$\Delta H = H_\delta - H_L = -15\ 190 \text{ J mol}^{-1}$$

This ΔH is called the *latent heat of solidification*. The negative sign tells us that when iron solidifies, latent heat is evolved from the iron, and this heat is available as a source of energy. If we think about melting (i.e., fusion), then

$$\Delta H = H_L - H_\delta = +15\ 190 \text{ J mol}^{-1}$$

and this ΔH is called the *latent heat of fusion* and given the symbol H_f.

Now suppose we fill a small ladle with liquid iron exactly at its freezing point (1809 K). Then there will be heat losses by convection and radiation from the top of the melt and by conduction through the wall of the ladle. When the total heat lost equals

Heat Transfer Fundamentals for Metal Casting
Second Edition with SI Units
Edited by D.R. Poirier and E.J. Poirier
The Minerals, Metals & Materials Society, 1994

30

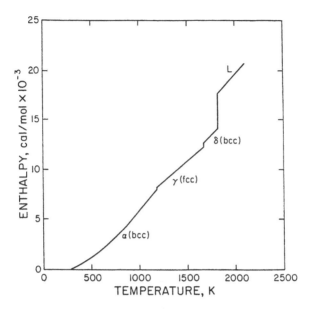

Figure 10: The enthalpy of iron. The conversion for energy is 1 cal = 4.1840 J.

the product of the mass and latent heat of fusion of the metal, then the iron in the ladle will be completely solidified.

The heat to be removed from the metal is

$$\hat{Q}_M = m H_f \tag{59}$$

where

$\hat{Q}_M$ = heat removed from metal, J

m = mass of metal, kg

H_f = latent heat of fusion, J kg^{-1}

The heat loss from the top surface is

$$\hat{Q}_T = Q_T t \tag{60}$$

where

$\hat{Q}_T$ = heat loss from top, J

Q_T = rate of heat loss from top, W

t = time, s

Q_T is by convection and radiation, so that

$$Q_T = (h + h_r) A_T (T_M - T_\infty) \qquad (61)$$

where

A_T = area of the top surface of metal, m^2

T_M = freezing point of the metal, K

T_∞ = temperature of the surroundings.

To get the total heat loss, we must also add the heat lost through the side wall of the ladle. To simplify our thought process, however, let's assume that the side wall of the ladle is a perfect insulator so that Eq. (61) accounts for all of the heat lost. Then, we merely make $\hat{Q}_M = \hat{Q}_T$, and substitute Eq. (61). Then we solve for the time required to solidify all of the iron. The result is

$$t = \frac{m H_f}{(h + h_r) A_T (T_M - T_\infty)}. \qquad (62)$$

Now consider a modification in the above scenario. The ladle is filled with superheated liquid iron, at a temperature T_P (i.e., the pouring temperature). In this case the heat to be removed from the metal comprises *sensible heat* and latent heat. Then

$$\hat{Q}_M = m C_{P,L} (T_P - T_M) + m H_f \qquad (63)$$

where $C_{P,L}$ is the specific heat of the liquid metal. Any specific heat is defined as

$$C_P = \frac{dH}{dT}$$

and is approximately constant over a range of temperature. For example, the slope of the enthalpy curve for liquid iron in Fig. 10 is constant so that $C_{P,L} = 791$ J K^{-1} kg^{-1}.

Again there is heat lost by convection and radiation from the top of the metal, but its temperature decreases from T_P to T_M as the liquid metal cools and then remains fixed at T_M as the liquid solidifies. For the first period t_1

$$Q_{Tl} \cong (h + h_r)_{\overline{T}} A_T (\overline{T} - T_\infty) \qquad (64)$$

where $\overline{T} = \frac{1}{2}(T_P + T_M)$ and the heat transfer coefficients are evaluated at $\overline{T}$. During this period the *superheat* is removed from the metal, so we can write

$$Q_{TI} t_1 = m\, C_{P,L} (T_P - T_M) \tag{65}$$

or

$$t_1 = \frac{m\, C_{P,L} (T_P - T_M)}{(h + h_r)_{\overline{T}}\, A_T (\overline{T} - T_\infty)}. \tag{66}$$

For the second period, we use Eq. (62) with t_2 substituted for t. Thus

$$t_2 = \frac{m\, H_f}{(h + h_r) A_T (T_M - T_\infty)} \tag{67}$$

and the total time is

$$t = t_1 + t_2. \tag{68}$$

Example 10

A ladle with 225 kg of liquid iron is initially at 1895 K. Estimate (i) the time required for the liquid iron to lose all of its superheat (i.e., sensible heat) and (ii) the total time for it to solidify. The exposed area is 0.12 m^2. Assume that the ladle wall is a perfect insulator.

For the first period, $\overline{T} = 1852$ K and we gather the properties for air at $T_f = \frac{1}{2}(1852 + 300) = 1076$ K from Table 4:

$\rho = 0.3305 \text{ kg m}^{-3}$

$\mu = 4.356 \times 10^{-5} \text{ kg m}^{-1} \text{ s}^{-1}$

$k = 0.0716 \text{ W m}^{-1} \text{ K}^{-1}$

$Pr = 0.704$

Also

$\beta = 1/T_f = 1/1076 = 9.29 \times 10^{-4} \text{ K}^{-1}$

$L \simeq \text{diameter} = 0.39 \text{ m}$

$g = 9.81 \text{ m s}^{-2}$

Then

$$Gr = \frac{(9.81)(9.29 \times 10^{-4})(0.3305)^2(1852-300)(0.39)^3}{(4.356 \times 10^{-5})^2}$$

$$= 4.83 \times 10^7$$

$$GrPr = 3.40 \times 10^7$$

and Table 7 gives $C = 0.15$ and $m = 1/3$.

$$Nu = \frac{hL}{k} = 0.15(3.40 \times 10^7)^{1/3} = 48.59$$

$$h = \frac{(48.59)(0.0716)}{(0.39)} = 8.92 \text{ W m}^{-2}\text{ K}^{-1}$$

The radiation heat transfer coefficient is evaluated with $\epsilon = 0.28$, $T_1 = 1852$ K and $T_2 = 300$ K. Then

$$h_r = (0.28)(5.670 \times 10^{-8})(1852^2 + 300^2)(1852 + 300)$$

$$= 120.26 \text{ W m}^{-2}\text{ K}^{-1}$$

Now Eq. (66) can be used, with $C_{P,L} = 791$ J kg^{-1} K^{-1} and $T_M = 1809$ K:

$$t_1 = \frac{(225)(791)(1895-1809)}{(8.92 + 120.26)(0.12)(1852-300)} = 636 \text{ s}$$

During the solidification period, Eq. (67) applies, but now the temperature is slightly less than $\overline{T}$ for the first period. The reduced temperature slightly changes the properties of air so that the convection heat transfer coefficient becomes $h = 8.93$ W m^{-2} K^{-1}. (To save space, this calculation is omitted.) The radiation heat transfer coefficient is evaluated with $T_1 = 1809$ K and $T_2 = 300$ K; the result is $h_r = 112.6$ W m^{-2} K^{-1}. Now Eq. (67) is applied with $H_f = 2.72 \times 10^5$ J kg^{-1}.

$$t_2 = \frac{(225)(2.72 \times 10^5)}{(8.91 + 112.6)(0.12)(1809-300)} = 2782 \text{ s}.$$

The total time is $t = t_1 + t_2 = 3417$ s (57.0 min). By far most of the time is while the iron solidifies.

It is instructive to plot temperature versus time when initially the liquid is at the freezing point and when the liquid has superheat. These cases are shown in Figs. 11a and b, respectively. In Fig. 11a, the temperature does not change while the metal solidifies. Only when solidification is complete does the temperature decrease. In Fig. 11b, temperature decreases as the liquid metal loses superheat, remains constant as latent heat is evolved during solidification, and then decreases as the solid cools. In Chapter IV, we apply these principles to the solidification of metal castings.

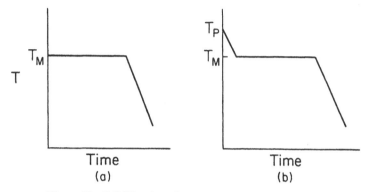

Figure 11: Solidification of a pure metal: (a) with no superheat and (b) with superheat.

B. *Solidification of Alloys*

Casting alloys solidify over a range of temperature so that the latent heat of fusion is not released at one temperature. Figure 12a illustrates the fraction solidified vs. temperature for an alloy that solidifies to a single phase. The enthalpy of the alloy during solidification, H, is

$$H = f_s H_s + f_L H_L$$

where

H_s, H_L = the enthalpy of the solid and liquid, respectively

and

f_s, f_L = the weight fraction of solid and liquid, respectively.

The enthalpy can also be written as

$$H = H_L - f_s H_f. \tag{69}$$

For the alloy shown in Fig. 12a, enthalpy vs. temperature is shown in Fig. 12b.

35

Figure 12c is drawn for an alloy that solidifies as a single phase with an eutectic constituent at the end of solidification. The eutectic solidifies at a fixed temperature (or over a small range). This accounts for the step equal to $f_E H_f$ in Fig. 12d, where f_E is the fraction of the eutectic constituent. The added complexity of the evolution of latent heat over a range of temperatures usually requires numerical computations for estimating temperatures and heat transfer during solidification.

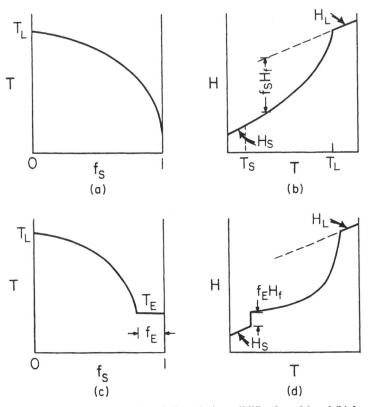

Figure 12: Fraction solid and enthalpy of alloys during solidification. (a) and (b) for an alloy that solidifies as one phase; (c) and (d) for an alloy that solidifies with an eutectic at the end of solidification.

III. HEAT TRANSFER IN MOLDS

A. Conduction in Mold Materials

Mold materials can be classified as insulators or as good conductors. The former include plaster full molds, ceramic shell molds in investment castings, and silica sand molds in sand castings. Permanent mold casting and die castings use metallic molds that are good conductors.

The conduction heat flux in a mold is expressed with the usual rate equation:

$$q = -k \frac{\partial T}{\partial x} \qquad (70)$$

The thermal conductivity of the mold, k, is an *effective* thermal conductivity, as contrasted to the *intrinsic* thermal conductivity of the individual particles that make up the mold. On a microscopic level, the mold is porous so heat is actually transferred by intrinsic conduction through each particle and conduction and convection in the gas in the pores. In addition, there is radiation from particle to particle by radiation across the pores. Therefore, the thermal conductivity, used in Eq. (70), is an effective conductivity because it depends on several factors: particle material, particle size, binder, volume fraction of pores, gas in the pores, emissivity of the particles, and temperature.

The effect of temperature on the thermal conductivity in insulating molds is shown in Fig. 13. The curves labeled with AFS sand numbers are for sand molds. In the sand molds, thermal conductivity increases with increasing temperature, because the fraction of energy transferred by radiation from particle to particle increases.

B. Heat Transfer across Casting Gaps

As an approximation, the temperature at the surface of an insulating mold adjacent to solidifying metal is equal to the freezing point of the metal. Because the thermal conductivity of the mold is only ~0.01 to 0.02 that of the metal, then practically all of the thermal resistance to the heat transfer is within the mold, itself. The temperature distribution in this situation is illustrated in Fig. 14a.

Early in the solidification process the resistance to heat transfer across the gap that forms at the mold-casting interface can also influence the overall heat transfer (Fig. 14b). If we knew precisely the characteristics of the gap, including its thickness variation with time, composition of the gas within the gap and the emissivities of the mold and casting surfaces, then the heat transfer across the gap could be estimated. However, such attempts usually fail so heat transfer associated with a gap is usually treated empirically with a *gap heat transfer coefficient* (h_g), which is defined by the flux across the gap. The flux across the gap is

$$q = h_g\left(T_c - T_M\right) = h_g \Delta T. \qquad (71)$$

Heat Transfer Fundamentals for Metal Casting
Second Edition with SI Units
Edited by D.R. Poirier and E.J. Poirier
The Minerals, Metals & Materials Society, 1994

37

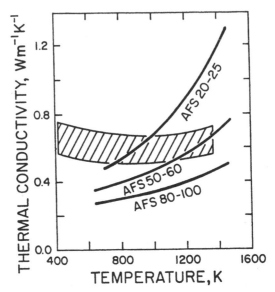

Figure 13: Thermal conductivity of insulating mold materials. Curves for various AFS numbers refer to sand molds. The cross hatched area is for ceramic shell molds with various binders.

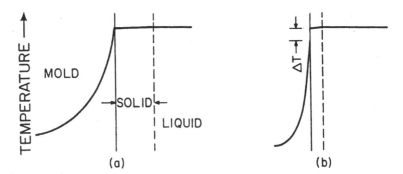

Figure 14: Temperature distributions when casting in an insulating mold: (a) with no thermal resistance at the casting-mold interface; (b) with thermal resistance at the casting mold interface.

where T_c and T_M are the temperatures at the casting surface and mold surface, respectively. Because the characteristics of the gap are not that well know, actual values of h_g can only be determined empirically.

Castings freeze relatively quickly in metal molds (i.e., conducting molds), and temperature changes drastically in both the mold and the casting. An understanding of the factors affecting solidification is important because permanent mold castings and die castings are made in metal molds. Also many castings are made in insulating molds with metal chills inserted at strategic positions to increase solidification rate.

The analysis of heat transfer when metal is poured against a chill is more complicated than when metal is poured against an insulating mold. The added complexities are illustrated in Fig. 15. At the metal-mold interface, the thermal resistance of the gap (or mold casting) causes a rather large temperature drop. The condition of no thermal resistance at this interface would exist only if the casting solders or welds to the mold, which obviously is not desirable.

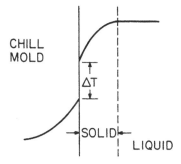

Figure 15: Temperature distribution when casting in a conducting mold with thermal resistance at the casting-mold interface.

In addition to the rather large temperature drop across the gap, there are other differences between solidification in insulating and chill molds:

(a) The thermal resistance of the metal being cast forms an important portion of the overall resistance to heat transfer.

(b) More total heat is removed during solidification because the solidified metal is appreciably below the melting point.

The value of ΔT across the gap depends upon the thermal properties of both the mold and the casting and the gap heat transfer coefficient. Gap heat transfer coefficients are given in Table 9.

Table 9

Gap Heat-Transfer Coefficients

Casting Situation	h_g $W\ m^{-2}\ K^{-1}$
Ductile iron in cast iron mold coated with amorphous carbon	1700
Steel in cast iron mold	1020
Aluminum alloy in small copper mold	1700-2550

Steel chilled by steel mold
    before gap forms*
    after gap forms
	400-1020
	400

Aluminum die castings
    before gap forms*
    after gap forms
	2500-5000
	400

*With no gap, thermal contact between the casting and the mold is not perfect because of surface tension effects, oxide layers and mold coatings.

IV. SOLIDIFICATION HEAT TRANSFER

The solidification rate of alloys is an important processing variable. For example, solidification rate relates directly to the coarseness (or fineness) of dendritic structures and hence controls the spacing and distribution of microheterogeneities, such as dendritic microsegregation, second phases, inclusions and microporosity. Although most types of macrosegregation are not directly related to solidification rate, it is known that larger castings (i.e., castings with relatively slower solidification rates) are more prone to macrosegregation than are smaller castings. For these metallurgical reasons and from a manufacturing viewpoint (e.g., production rate and riser design), metal casting engineers should recognize solidification heat transfer as an important topic.

A. Solidification in Thick Molds

Consider pure liquid metal with no superheat poured against a flat surface of an insulating mold (Fig. 14a). We make the following assumptions:

(1) no thermal resistance at the casting-mold interface;

(2) because all of the resistance to heat transfer is almost entirely within the mold, there is no temperature gradient in the solidified portion of the casting;

(3) the mold is semi-infinite in extent and its thermal properties are constant and uniform.

By assumptions 1 and 2, the temperature at the surface of the mold $(x = 0)$ is equal to the freezing point of the metal (T_M). Then with assumption 3, Eq. (31) applies so the rate of heat flow into the mold is

$$Q = \frac{A}{\sqrt{\pi}} \sqrt{k\rho C_p} \ (T_M - T_i) \, t^{-1/2} \qquad (72)$$

where A is the area of the mold-casting interface and T_i is the initial temperature of the mold. The product $k\rho C_p$ represents the ability of the mold to absorb heat at a certain rate. It is called *heat diffusivity* and is not to be confused with the *thermal diffusivity* $(\alpha = k/\rho C_p)$.

With ρ' as the density of the solid metal and M as the thickness solidified, $\rho' A dM/dt$ represents the rate of increase of solid (kg s^{-1}). If it is multiplied by the latent heat we get the heat extracted from the casting:

$$Q = \rho' H_f A \, \frac{dM}{dt} \qquad (73)$$

Then Eqs. (72) and (73) are set equal; this gives

$$\frac{dM}{dt} = \frac{1}{\sqrt{\pi}} \left[\frac{T_M - T_i}{\rho' H_f} \right] \sqrt{k\rho C_p} \ t^{-1/2} \qquad (74)$$

Heat Transfer Fundamentals for Metal Casting
Second Edition with SI Units
Edited by D.R. Poirier and E.J. Poirier
The Minerals, Metals & Materials Society, 1994

41

Integration of Eq. (74) follows with $M = 0$ at $t = 0$; the result is

$$M = \frac{2}{\sqrt{\pi}} \left[\frac{T_M - T_i}{\rho' H_f} \right] \sqrt{k\rho C_p} \; t^{1/2}. \tag{75}$$

We see that the amount of solidification depends on the properties of the metal (the quantities in the parentheses) and the mold's heat diffusivity. Some mold and metal data are given in Table 10. .

Table 10

Properties for Insulating Molds and Solidifying Metals

Mold Material*	k W m^{-1} K^{-1}	ρ kg m^{-3}	C_p J kg^{-1} K^{-1}	$\sqrt{k\rho C_p}$ J m^{-2} K^{-1} s$^{-1/2}$
Silica Sand	0.52	1600	1170	987
Mullite	0.38	1600	750	675
Plaster	0.35	1120	840	574
Zircon Sand	1.04	2720	840	1540

Casting Material	T_M K	H_f J kg^{-1}	ρ' kg m^{-3}	C_p' J kg^{-1} K^{-1}	k' W m^{-1} K^{-1}
Iron	1809	2.72×10^5	7210	750	40
Nickel	1728	2.91×10^5	7850	670	35
Aluminum	933	3.91×10^5	2400	1050	260

*Only typical values can be given here. Actual properties depend on temperature, particle size, binders, etc.

Example 11

Estimate the solidification time for iron (12.7 mm thick) cast in a zircon sand mold.

The metal solidifies from both faces of the mold. Thus,

$$M = 6.35 \text{ mm} = 0.00635 \text{ m}$$

Solving Eq. (75) for t:

$$t = \frac{\pi}{4} M^2 \left[\frac{\rho' H_f}{T_M - T_i} \right]^2 (k\rho C_p)^{-1}$$

$$t = \left[\frac{\pi}{4} \right] (0.00635^2) \left[\frac{7210 \times 2.72 \times 10^5}{1809 - 300} \right]^2 (1540^2)^{-1} = 22.6 \ s$$

Now we consider the effect of thermal resistance at the casting-mold interface with an approximate analysis. Assume that the temperature in the solidified metal is linear. Then the thermal resistance in the solid metal is M/Ak'. By making the analogy between Ohm's law and Eq. (72), the thermal resistance in the mold is

$$R_T = \frac{\sqrt{\pi t}}{A \sqrt{k\rho C_p}} . \tag{76}$$

With the thermal resistance of the gap taken as $1/Ah_g$, we get the total thermal resistance

$$R_T = \frac{M}{Ak'} + \frac{1}{Ah_g} + \frac{\sqrt{\pi t}}{A \sqrt{k\rho C_p}} . \tag{77}$$

Then Eq. (72) can be replaced with

$$Q = \frac{T_M - T_i}{R_T}$$

or

$$Q = \frac{A(T_M - T_i)}{\left[\frac{M}{k'} + \frac{1}{h_g} + \left[\frac{\pi t}{k\rho C_p} \right]^{1/2} \right]} . \tag{78}$$

To keep the problem tractable, we ignore the sensible heat extracted from the solid and make Eqs. (73) and (78) equal. Further we recognize that $1/h_g \gg M/k'$ and neglect M/k'. The result is

$$\frac{dM}{dt} = \left[\frac{T_M - T_i}{\rho' H_f} \right] \left[\frac{1}{h_g} + \frac{\sqrt{\pi t}}{\sqrt{k\rho C_p}} \right]^{-1} . \tag{79}$$

Equation (79) is integrated with $M = 0$ at $t = 0$; the result is

$$M = \frac{2}{\sqrt{\pi}} \left[\frac{T_M - T_i}{\rho' H_f} \right] \sqrt{k\rho C_p} \; \sqrt{t} \left\{ 1 - \frac{\sqrt{k\rho C_p}}{h_g \sqrt{\pi t}} \ln \left[1 + \frac{h_g \sqrt{\pi t}}{\sqrt{k\rho C_p}} \right] \right\} \quad (80)$$

The term in brackets can be looked upon as a factor, $\phi(t)$, which accounts for resistance to heat transfer because of the gap. Figure 16 is a plot of ϕ vs. time for plaster and zircon sand molds (Table 10) with $h_g = 570$ and 2840 W m^{-2} K^{-1}. For small times (i.e., thin section castings), the gap retards solidification significantly. The retarding effect is more in zircon sand molds than in plaster molds because the heat diffusivity of zircon sand is greater than that of plaster.

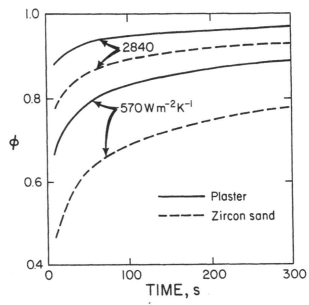

Figure 16: ϕ for various values of the gap heat transfer coefficient in insulating molds.

Example 12

Repeat the problem of *Example 11* and assume that $h_g = 2840$ W m^{-2} K^{-1}.

By our previous calculation, an estimate of the solidification time is 23 s. For 23 s, Fig. 16 shows $\phi \approx 0.82$. Inspection of *Example 11* shows that $\phi^2 t = 23$ s.

Thus, $t = \dfrac{23}{(0.82)^2} = 34$ s

B. *Modulus Method*

For metal castings made in insulating molds, the modulus is defined as the volume to surface area ratio, V/A. The volume V can be the entire volume of the casting or some portion of the casting, and A is the corresponding casting-mold interfacial area. For infinite plates, $V/A = M$ where M is the semithickness. Then we can write Eq. (80) as

$$\frac{V}{A} = \frac{2\phi}{\sqrt{\pi}} \left[\frac{T_M - T_i}{\rho' H_f} \right] \sqrt{k\rho C_p} \sqrt{t} \tag{81}$$

where t is the solidification time of the casting and an appropriate value of ϕ is used. This relationship is solved for t with result:

$$t_s = C \left[\frac{V}{A} \right]^2 \tag{82}$$

where

$$C = \frac{\pi}{4\phi^2} \left[\frac{\rho' H_f}{T_M - T_i} \right]^2 \left[\frac{1}{k\rho C_p} \right].$$

Equation (82) is often referred to as Chvorinov's role, and C, as Chvorinov's constant. It is intended for comparison of freezing times of castings with different shapes and sizes. The relationship works best for casting geometries in which none of the mold material becomes saturated with heat, such as in internal corners or internal cores. Its success hinges on the mold material absorbing the same amount of heat for every unit area exposed to the casting. This is strictly true only for a group of castings, which have similar geometries but different sizes.

To quantify effects of various geometries of castings, let us examine differences among three basic shapes: namely infinite plates, infinite circular cylinders, and spheres. First we define two dimensionless parameters, β and γ:

$$\beta = \frac{(V/A)}{\sqrt{\alpha t}} \tag{83}$$

and

$$\gamma = \phi \left[\frac{T_M - T_i}{\rho' H_f} \right] \rho C_p. \tag{84}$$

Then the freezing times for the three basic shapes are given by

$$\beta = \gamma \left[\frac{2}{\sqrt{\pi}} + \frac{1}{a\beta} \right] \tag{85}$$

45

where $a = \infty$ for infinite plates, $a \simeq 4$ for infinite circular cylinders and $a = 3$ for spheres.

We can deduce Eq. (85) from Eq. (81) for infinite plates, with $a = \infty$. Likewise Eq. (85) can be deduced from available equations for cylinders and spheres, although those equations are not given here.

Example 13

Compare the freezing time of an iron sphere, 50 mm diameter, to a plate that is 50 mm thick. Both are cast in a zircon sand mold. Assume that $\phi = 0.9$.

First Eq. (84) is used to estimate γ with thermal properties from Table 10.

$$\gamma = 0.9 \left[\frac{1809 - 300}{7210 \times 2.72 \times 10^5} \right] (2720 \times 840) = 1.582.$$

Then Eq. (85) is used to get β for both the plate and the sphere.

$$\beta_p = \beta(\text{plate}) = 1.582 \left[\frac{2}{\sqrt{\pi}} \right]$$

$$\beta_s = \beta(\text{sphere}) = 1.582 \left[\frac{2}{\sqrt{\pi}} + \frac{1}{3\beta_s} \right].$$

Solving:

$$\beta_p = 1.78 \quad \text{and} \quad \beta_s = 2.04.$$

For the plate,

$$\frac{V}{A} = 25 \text{ mm} = 0.025 \text{ m}$$

$$\alpha = \frac{k}{\rho C_p} = \frac{1.04}{(2720)(840)} = 4.55 \times 10^{-7} \text{ m}^2 \text{ s}^{-1}$$

$$t(\text{plate}) = \frac{(V/A)^2}{\beta_p^2 \alpha} = \frac{0.025^2}{(1.78^2)(4.55 \times 10^{-7})}$$

$$t(\text{plate}) = 434 \text{ s} = 7.23 \text{ min.}$$

For the sphere (d = dia.),

$$\frac{V}{A} = \frac{d}{6} = 0.00833 \text{ m}$$

$$t(\text{sphere}) = \frac{0.00833^2}{(2.04^2)(4.55 \times 10^{-7})}$$

$$t(\text{sphere}) = 36.6 \text{ s } (0.6 \text{ min.}) \quad .$$

The above example clearly illustrates that castings of the same thickness do not solidify in the same time. The modulus should be used as the characteristic dimension of the casting when solidification times are estimated. Furthermore, shapes that favor *divergent* heat flux lines in the mold solidify in less time than shapes made in molds with unidirectional heat flux lines or *convergent* heat flux lines.

Example 14

From *Example 13* the solidification time of a 50 mm dia. sphere is 36.6 s. Estimate the freezing time of a 82.5 mm cube.

Using subscripts s and c for the sphere and cube, respectively, Chvorinov's rule gives

For the sphere:
$$\frac{t_c}{t_s} = \frac{(V/A)_c^2}{(V/A)_s^2}$$

$$\left[\frac{V}{A}\right]_s = \frac{d}{6} = 8.333 \text{ mm}$$

For the cube, with ℓ as the cube edge length:

$$\left[\frac{V}{A}\right]_c = \frac{\ell}{6} = 13.75 \text{ mm}$$

Then

$$t_c = \left[\frac{13.75}{8.333}\right]^2 (36.6) = 100 \text{ s}$$

(Notice that we have assumed $a = 3$, in Eq. (85), for the cube.)

We can approximate the effect of superheat on solidification time by realizing that, in addition to absorbing latent heat, the mold must also absorb superheat. The total quantity of heat to be removed from the casting is

$$\hat{Q} = \rho' V H_f + \rho'_\ell V C_{p,\ell} \Delta T_s. \tag{86}$$

The subscript ℓ denotes liquid phase properties, and ΔT_s is the amount of superheat. In order to make the analysis simple, yet sufficiently accurate, we assume that the interface temperature of the mold is constant while the liquid loses its superheat. In addition, $\rho'_\ell \simeq \rho'$ then

$$\hat{Q} = \rho' V H'_f \tag{87}$$

where H'_f is the *effective heat of fusion*. That is

$$H'_f = H_f + C_{p,\ell} \Delta T_s$$

and H'_f replaces H_f in all of the previous equations that are used to estimate solidification time. Note that solidification time is still proportional to $(V/A)^2$.

C. *Solidification in Molds of Finite Thickness*

In this section we consider cases in which the mold cannot be assumed to be infinitely thick. Figure 17 illustrates a situation that often leads to a hot spot in a casting. Part of the mold or a core is surrounded by the solidifying metal. If this part of the mold is thin enough, it can become saturated with heat and incapable of extracting further energy from the solidifying metal.

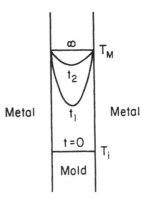

Figure 17:   Temperature distribution in a mold or core surrounded by the solidifying metal.

48

Heat conduction in the mold must satisfy Eq. (10) and

initial temperature: $T(x,0) = T_i$ (uniform)

surface $(x = L)$: $T(L,t) = T_M$

centerline $(x = 0)$: $\dfrac{\partial T}{\partial x}(0,t) = 0$.

The particular solution that satisfies these conditions can be found in many books on conduction heat transfer. Because the mold has a finite thickness, the solution is in the form of an infinite series, which is not repeated here. Instead we write the equation for the heat absorbed by the mold, which can be derived from the infinite series for the temperature distribution. The equation is

$$\frac{\hat{Q}_M/A}{\rho C_p L\left(T_M - T_i\right)} = 2 \sum_{n=0}^{\infty} \beta_n^{-2}\left[1 - \exp\left(-\beta_n^2 F_0\right)\right] \tag{89}$$

where

$$\beta_n \equiv (2n + 1)\pi/2$$

$$Fo = \frac{\alpha t}{L^2}.$$

If A is taken to be the area of one face of the mold portion, then $\hat{Q}_M$ is the energy absorbed by one half of the mold. If A represents the area of both faces, then $\hat{Q}_M$ is the energy absorbed by the whole mold portion. The Fourier number, Fo, is dimensionless time and appears in conduction problems.

As usual the energy extracted from the solidified casting is $\rho' V H_f'$, and it is substituted for $\hat{Q}_M$. After rearranging, Eq. (89) is written in dimensionless form as

$$\frac{V/A}{L} = 2\gamma \sum_{n=0}^{\infty} \beta_n^{-2}\left[1 - \exp\left(-\beta_n^2 Fo\right)\right] \tag{90}$$

where γ is defined by Eq. (84).

Equation (90) is plotted in Fig. 18 for typical values of γ. Each curve in Fig. 18 rises to a plateau when the mold is saturated with heat, and there is no additional solidification.

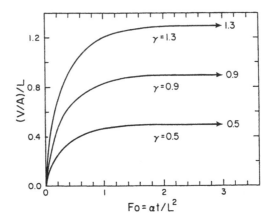

Figure 18: Solidification against a mold or core surrounded by metal.

Example 15

Estimate the maximum thickness of aluminum that is solidified from both sides of a plaster mold that is 10 mm thick. The mold is preheated to 480 K

First we calculate γ with properties from Table 10 (assume that $\phi = 1$).

$$\gamma = (1120)(840) \left[\frac{933 - 480}{2400 \times 3.91 \times 10^5} \right] = 0.454$$

The limiting values of $(V/A)/L$ are equal to γ (see Fig. 18); this corresponds to the situation when the mold is completely saturated with heat and $t = \infty$ in Fig. 17. Then

$$\frac{(V/A)}{L} = 0.454.$$

Finally, with $L = (1/2)(10 \text{ mm}) = 5$ mm, the thickness solidified is

$$M = V/A = (0.454)(5) = 2.27 \text{ mm}$$

This simple calculation shows that a mold or core surrounded by metal can easily become saturated with heat.

Now we consider solidification in a shell mold with a thickness L and heat loss from the outside surface to the surroundings. Again Eq. (10) applies. The initial condition and the boundary conditions are

$$\text{initial temperature:} \quad T(x,0) = T_i \text{ (uniform)}$$

$$\text{mold-metal surface } (x = 0): \quad T(0,t) = T_M$$

$$\text{mold-surroundings surface } (x = L): \quad \frac{\partial T}{\partial x}(L,t) + \frac{h}{k}\left[T(L,t) - T_\infty\right] = 0.$$

The last condition is derived by equating the conduction heat flux in the mold to the flux of heat lost to the surroundings.

The method of solution to this particular set of conditions is rather lengthy so only the final result is given. With dimensionless groups, it is

$$\frac{V/A}{L\gamma'} = \left\{\frac{BiFo}{1 + Bi}\right\}\left\{1 + \frac{2}{Fo}\sum_{n=1}^{\infty}\frac{B_n}{w_n}\frac{\left(Bi^2 + w_n^2\right)}{Bi^2 + w_n^2 + Bi}\exp\left(-w_n^2 Fo\right)\right\} \quad (91)$$

where the w_ns are the positive roots of

$$w_n \cot w_n + Bi = 0, \quad (92)$$

the B_ns are given by

$$B_n = \left[\frac{T_M - T_i}{T_M - T_\infty}\right]\left[\frac{1 + Bi}{Bi}\right]\left[\frac{1 - \cos w_n}{w_n}\right] + \left[\frac{\cos w_n}{w_n} - \frac{\sin w_n}{w_n^2}\right], \quad (93)$$

and

$$\gamma' = \rho C_p \left[\frac{T_M - T_\infty}{\rho' H_f'}\right].$$

Examination of Eqs. (91)-(93) shows that there are four independent dimensionless groups:

(1) $(V/A)/L\gamma'$

(2) Bi (Biot number) $= hL/k$

(3) Fo (Fourier number) $= \alpha t/L^2$

(4) $(T_M - T_i)/(T_M - T_\infty)$.

51

The first and third dimensionless numbers give the modulus of the casting versus solidification time. Properties of the metal and mold are evident in the first, second and third dimensionless numbers. Finally the effects of the mold thickness, mold preheat, and the heat transfer coefficient for the heat loss to the surroundings are also included in the entire set.

Results for solidification in molds with preheat and with no preheat are shown in Fig. 19. Except for relatively short times, the results depend mainly on the Biot number and are almost independent of the preheat. This can be seen in Fig. 19a. The differences between preheat and no preheat are more apparent for short times (Figs. 19b and 19c).

Provided $Fo \geq 1$, the results can be represented as

$$\frac{V/A}{L} = \gamma' \left[\frac{Bi}{1 + Bi} \right] Fo + b \tag{94}$$

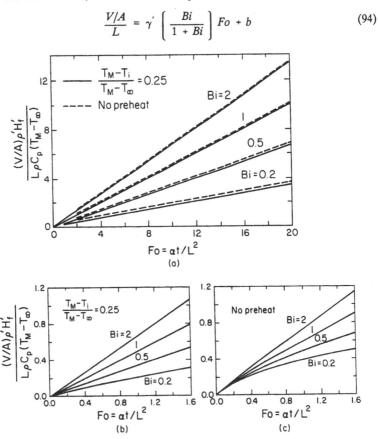

Figure 19: Solidification in molds of finite thickness; (a) long times; (b) short times with preheat; (c) short times with no preheat.

where b is an intercept. In the linear part of the curves, the temperature distribution in the mold is very close to steady state as the casting solidifies. With no preheat, more time is required to achieve steady state; this can be seen by examination of Figs. 19b and 19c. The value of the intercept depends on Bi and the preheat; in all cases $b \leq 0.25$.

Example 16

Estimate the solidification time for nickel cast with 55 K superheat in a ceramic shell mold (7.5 mm thick) preheated to 1250 K. Assume that the properties of mullite in Table 10 are appropriate for the shell mold, and that $h = 100$ W m^{-2} K^{-1}. The modulus of the casting is 25.4 mm.

Assume that the effective latent heat of fusion can be used to account for both the latent heat and the superheat. Then $H_f' = 2.91 \times 10^5 + 670(55) = 3.28 \times 10^5$ J kg^{-1}. Equations (91)-(93) are best solved with a computer code because Eq. (91) contains an infinite series and Eq. (92) cannot be solved explicitly. Evaluation of the dimensionless numbers is as follows:

$$Bi = \frac{hL}{k} = \frac{(100)(0.0075)}{(0.38)} = 1.97$$

$$\frac{T_M - T_i}{T_M - T_\infty} = \frac{1728 - 1250}{1728 - 300} = 0.335$$

$$\gamma' = \rho C_p \left[\frac{T_M - T_\infty}{\rho' H_f'} \right] = (1600)(750) \left[\frac{1728 - 300}{(7850)(3.28 \times 10^5)} \right] = 0.666$$

$$\frac{(V/A)}{L\gamma'} = \frac{(0.0254)}{(0.0075)(0.666)} = 5.08$$

$$Fo = \frac{kt}{\rho C_p L^2} = \frac{0.38\ t}{(1600)(750)(0.0075)^2} = 5.63 \times 10^{-3}\ t \ (t \text{ in s})$$

Equations (91)-(93) give the following results:

Fo	$(V/A)/L\gamma'$
0.335	0.219
2.011	1.330
3.686	2.442
5.362	3.552
7.037	4.664
8.713	5.776
10.388	6.888

Interpolation gives $Fo = 7.66$, so that

$$t = \frac{7.66}{5.63 \times 10^{-3}} = 1360 \text{ s } (22.7 \text{ min.}) \ .$$

In *Example 16*, $b \simeq 0$ so that Eq. (94) could be approximated as the same result obtained if the temperature in the mold had been assumed to be at steady state. This result is

$$\frac{V}{A} = Ct \tag{95}$$

where the constant C is defined as

$$C \equiv \frac{k}{L} \left[\frac{T_M - T_\infty}{\rho' H_f'} \right] \left[\frac{Bi}{1 + Bi} \right] \ . \tag{96}$$

Now suppose we consider solidification in a cylindrical shell mold, with the mold in the domain $R_1 \leq r \leq R_2$. At the casting mold interface ($r = R_1$), the temperature is T_M, and at the outer surface of the mold ($r = R_2$) there is heat loss to the surroundings at T_∞. It can be shown that the steady state heat lost through the inside area of the mold (A) in time t is

$$\hat{Q}_m = \frac{Ah(T_M - T_\infty)t}{(R_1/R_2) + (hR_1/k) \ln (R_2/R_1)} \tag{97}$$

As usual

$$\hat{Q}_c = \rho' V H_f' \tag{98}$$

so that with $\hat{Q}_M = \hat{Q}_c$ and $t = t_s$ (solidification time), we get

$$\frac{V}{A} = Ct \tag{99}$$

where

$$C \equiv \left[\frac{k}{R_2 - R_1} \right] \left[\frac{T_M - T_\infty}{\rho' H_f'} \right] \left[\frac{Bi}{(R_1/R_2) + R_1 \ln (R_2/R_1) Bi/(R_2 - R_1)} \right]$$

and

$$Bi \equiv \frac{h(R_2 - R_1)}{k} \ . \tag{100}$$

Example 17

Compare the values of C for the solidification of a plate and of a cylinder in shell molds of the same thickness. Use the same conditions of *Example 16*.

$$H_f' = H_f + C_p'(T_p - T_M) = 3.28 \times 10^5 \text{ J kg}^{-1}.$$

For the plate, we substitute values into Eq. (96), with $L = 0.0075$ m.

$$Bi = \frac{hL}{k} = \frac{(100)(0.0075)}{0.38} = 1.97$$

$$C = \left[\frac{0.38}{0.0075} \right] \left[\frac{1728 - 300}{7850 \times 3.28 \times 10^5} \right] \left[\frac{1.97}{2.97} \right] = 1.86 \times 10^{-5} \text{ m s}^{-1} \quad .$$

For the cylinder, we assume the same modulus. Then

$$\frac{V}{A} = 25.4 \text{ mm} = \frac{R_1}{2} ,$$

$$R_1 = 50.8 \text{ mm}$$

and for same thickness of mold, $R_2 - R_1 = 7.5$ mm. Now substituting these values into Eq. (100) we get

$$C = \left[\frac{0.38}{0.0075} \right] \left[\frac{1728 - 300}{7850 \times 3.28 \times 10^5} \right] \left[\frac{1.97}{0.8714 + (0.0508)(\ln 1.148)(1.97)/(0.0075)} \right]$$

$$C = 2.04 \times 10^{-5} \text{ m s}^{-1} \quad .$$

For the same modulus, the cylinder solidifies in about 10% less time than does the plate.

V. RISER DESIGN

A. *Riser Size*

We have seen that the solidification time of a simple shape cast in a thick mold of an insulating material can be estimated with Chvorinov's rule, which is

$$t = C(V/A)^2 \tag{82}$$

where

t = solidification time of the casting,

V = volume of the casting,

A = surface area of the casting,

C = Chvorinov's constant.

Let's first consider the case when Chvorinov's constants of the casting and riser are equal. The most efficient riser from freezing time point of view is the one in which solidification ceases simultaneously with the solidification in the casting. Because the freezing times are equal for the riser and the casting, Chvorinov's rule requires that

$$\frac{V_{Rf}}{A_R} = \frac{V_c}{A_c} \tag{101}$$

where

V_{Rf} = final volume of the riser (after complete solidification),

V_c = volume of the casting,

A_R = surface area of the riser, and

A_c = surface area of the casting.

Conceptually, volumes are proportional to the amount of heat that must be removed for complete solidification, and areas are proportional to the amount of heat that can be absorbed by the mold or extracted through the mold. Notice that the final volume of the riser is used because, this volume (and not its initial volume) is proportional to the heat extracted from the riser through area A_R.

A material balance on a riser-casting system, with all the shrinkages resulting in the riser, is

$$V_{Rf} = V_R - f(V_R + V_c) \tag{102}$$

Heat Transfer Fundamentals for Metal Casting
Second Edition with SI Units
Edited by D.R. Poirier and E.J. Poirier
The Minerals, Metals & Materials Society, 1994

56

in which

V_R = initial volume of the riser, and

f = fraction of feed metal requirement.

The quantity $f(V_R + V_c)$ is simply the total feed metal required to feed both the casting (V_c) and the riser (V_R).

By combining Eqs. (101) and (102), the basic riser equation is obtained; it is

$$\frac{A_R}{A_c} = (1 - f) \frac{V_R}{V_c} - f \tag{103}$$

The feed metal requirement is a result of metal shrinkage and mold dilation, and as such f can be process dependent, depending upon the degree of mold dilation. If necessary, it can be measured empirically by titrating water into the shrinkage cavities of risers during the development stage of a production run.

Recalling that the solidification times of simple shapes can be represented as

$$\beta = \gamma \left[\frac{2}{\sqrt{\pi}} + \frac{1}{a\beta} \right] \tag{85}$$

we can replace Eq. (101) with

$$\frac{V_{Rf}/A_R}{\beta_R \sqrt{\alpha_R}} = \frac{(V/A)_c}{\beta_c \sqrt{\alpha_c}} \tag{104}$$

$(V/A)_c$ is the volume-to-surface area ratio of the casting or the portion thereof fed by the riser, and subscripts R and c refer to the properties associated with the riser and the casting, respectively. Then by combining Eqs. (102) and (104), we get

$$(1 - f)V_R - fV_c = \left[\frac{\beta_R \sqrt{\alpha_R}}{\beta_c \sqrt{\alpha_c}} A_R \right] (V/A)_c \tag{105}$$

This equation is more general than Eq. (103) because it allows us to consider cases when the casting and riser have different geometries and/or different molding materials surrounding them. However, its use presumes that the areas of each are uniformly molded with its particular mold material.

Example 18

An aluminum plate casting, 25 mm × 125 mm × 125 mm, is fed with an end riser; both are entirely encased in silica sand. Calculate the height and diameter of the riser.

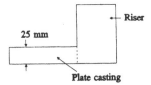

Neglect the contact area between the casting and the riser; this adds some conservatism to our analysis. The edges and corners of the plate will freeze more quickly than the plate as a whole, but the riser must remain molten long enough so that it does not freeze before the entire casting. Therefore, $(V/A)_c$ should be selected as that of an infinite plate. Then

$$\left[\frac{V}{A} \right]_c = 0.0125 \text{ m (i.e., semithickness)}$$

The shrinkage of pure aluminum is 0.06; to be safe we add a fraction for mold dilation and assume that $f = 0.075$. The volume of the casting is $V_c = 3.906 \times 10^{-4} \text{ m}^3$

Now we must estimate the values of the βs in Eq. (105). First consider the plate (Table 10):

$$\gamma = (1600)(1170) \left[\frac{933 - 300}{2400 \times 3.91 \times 10^5} \right] = 1.26$$

Then (with $a = \infty$ for plates), Eq. (85) gives

$$\beta_c = \frac{(2)(1.26)}{\sqrt{\pi}} = 1.42.$$

Now consider the riser. It is certainly not an infinitely long riser, so we can expect that $3 \leq a \leq 4$. For our estimate, let's assume that $a = 3.5$. Then

$$\beta_R = 1.26 \left[\frac{2}{\sqrt{\pi}} + \frac{1}{3.5\,\beta_R} \right],$$

from which, we find $\beta_R = 1.64$. Now with $\alpha_R = \alpha_c$, we have

$$\frac{\beta_R \sqrt{\alpha_R}}{\beta_c \sqrt{\alpha_c}} = \frac{1.64}{1.42} = 1.16.$$

Let's summarize Eq. (105), by inserting known values up to this point:

$$0.925 \, V_R - (0.075)(3.906 \times 10^{-4}) = (1.16)(0.0125) A_R.$$

For the cylindrical riser (D = dia. and H = hgt.),

$$V_R = \frac{\pi D^2 H}{4}$$

and

$$A_R = A(\text{top}) + A(\text{bottom}) + A(\text{side})$$

$$= 2 \left[\frac{\pi D^2}{4} \right] + \pi D H.$$

We are left with two unknown dimensions, D and H. The minimum diameter of the riser corresponds to a riser that is infinitely tall. Of course, this is not a practical solution, but it is interesting to determine the minimum diameter. As $H \to \infty$, $fV_c \to 0$ and $(V_R/A_R) \to D/4$. Then

$$D_{min} = 4 \left[\frac{1.16}{0.925} \right] (0.0125) = 0.0627 \text{ m} = 62.7 \text{ mm}$$

To obtain a practical result, let's assume that $H = D$. Then

$$V_R = \frac{\pi}{4} D^3, \qquad A_R = \frac{3\pi}{2} D^2,$$

and

$$(0.925) \left[\frac{\pi}{4} \right] D^3 - (1.16)(0.0125) \left[\frac{3\pi}{2} \right] D^2 = (0.075)(3.906 \times 10^{-4})$$

We have a choice: either solve the cubic equation directly or solve it by trial and error. The latter is easier, particularly because we already know D_{min}. The solution is $D = H = 98.2$ mm.

This might seem like a rather large riser, but we should keep in mind that a pure metal, in fact, exhibits piping. Alloys that freeze in a mushy manner, do not pipe very much and the shrinkage takes the form of microporosity dispersed in both the riser and the casting. Hence, less feed metal is required.

Example 19

Repeat *Example 18* for a mushy freezing aluminum alloy ($f = 0.04$) and select a riser with $H/D = 1$.

By following the solution to *Example 18*, but using $f = 0.04$, we get

$$0.960\,V_R - (0.04)(3.906 \times 10^{-4}) = (1.15)(0.0125)\,A_R$$

Now we substitute $H = D$, so that

$$V_R = \frac{\pi D^3}{4} \quad \text{and} \quad A_R = \frac{3\pi D^2}{2}.$$

Then

$$(0.960)\left[\frac{\pi}{4}\right] D^3 - (1.16)(0.0125)\left[\frac{3\pi}{2}\right] D^2 = (0.04)(3.906 \times 10^{-4}).$$

The solution to the cubic equation is $D = H = 93.0$ mm.

When an insulating sleeve is used around the riser, the riser is in contact with different materials on its sides, top and bottom. A top riser can be in contact with insulating material on its sides, with insulating, exothermic or sand material on its top and with the casting on its bottom. A side riser on the other hand, can be in contact with similar materials on its sides and top, but it can also be in contact with sand or insulating material on its bottom. To account for these various possibilities, therefore, we split the riser into three components for its side, top and bottom, then

$$\beta_R \sqrt{\alpha_R}\ A_R = \beta_t \sqrt{\alpha_t}\ A_t + \beta_b \sqrt{\alpha_b}\ A_b + \beta s \sqrt{\alpha_s}\ A_s. \tag{106}$$

The subscripts represent the top (t), bottom (b), and side (s) of the riser. By combining Eqs. (105) and (106), we arrive at

$$(1 - f)V_R - fV_c = \left[eA_s + z_1 A_t + z_2 A_b\right](V/A)_c \tag{107}$$

where

$$e = \frac{\beta_s \sqrt{\alpha_s}}{\beta_c \sqrt{\alpha_c}}\ \text{the \textit{apparent surface alteration factor} (asa) for the sides of the riser,}$$

$$z_1 = \frac{\beta_t \sqrt{\alpha_t}}{\beta_c \sqrt{\alpha_c}}\ \text{the asa factor for top of the riser,}$$

and

$$z_2 = \frac{\beta_b \sqrt{\alpha_b}}{\beta_c \sqrt{\alpha_c}}$$ the asa factor for bottom of the riser.

Let us define $\chi = H/D$ for the riser, with H its height and D its diameter. Now,

$$V_R = \frac{\pi}{4} D^3 \chi$$

$$A_s = \pi D^2 \chi \qquad (108a,b,c)$$

and

$$A_t = A_b = \frac{\pi}{4} D^2$$

Substituting Eqs. (108a,b,c) into Eq. (107) we arrive at the riser design equation:

$$D^3 - K_1(4e\chi + z_1 + z_2)D^2 - K_2 = 0 \qquad (109)$$

with the constants K_1 and K_2 defined as

$$K_1 = \frac{(V/A)_c}{(1 - f)}$$

$$K_2 = \frac{4fV_c}{\pi(1 - f)}$$

Example 20

An aluminum bar casting, 50 mm × 50 mm × 125 mm, is fed with a top riser that is encased in an insulating sleeve with a top. Assume that $f = 0.04$ and $H/D = 2$. The sleeve has the following properties: $k = 0.12$ W m^{-1} K^{-1}; $\rho = 230$ kg m^{-1} ; $C_p = 1090$ J kg^{-1} K^{-1}.

$$\left[\frac{V}{A} \right]_c = 0.0125 \text{ m} \qquad \text{neglecting the ends}$$

$$\gamma \text{ (casting)} = 1.26 \qquad \text{(from } Example\ 18\text{)}$$

61

As a long square bar, the casting is approximated as a cylinder. Therefore, $a = 4$ and

$$\beta_c = 1.26 \left[\frac{2}{\sqrt{\pi}} + \frac{1}{4\beta_c} \right] ;$$

solving $\beta_c = 1.62$.

Now we deal with the riser (side and top are encased in the insulator).

$$\gamma(\text{riser}) = (230)(1090) \left[\frac{933-300}{2400 \times 3.91 \times 10^5} \right] = 0.169$$

With $a \approx 3.5$,

$$\beta_R = 0.169 \left[\frac{2}{\sqrt{\pi}} + \frac{1}{3.5\,\beta_R} \right]$$

and solving $\beta_R = 0.335$.

The thermal diffusivities for the materials in contact with the casting and the riser are:

$$\alpha_c = \frac{(0.52)}{(1600)(1170)} = 2.78 \times 10^{-7} \text{ m}^2 \text{ s}^{-1}$$

and

$$\alpha_R = \frac{(0.12)}{(230)(1090)} = 4.79 \times 10^{-7} \text{ m}^2 \text{ s}^{-1}$$

The asa factors are

$$e = z_1 = \frac{0.335\sqrt{4.79 \times 10^{-7}}}{1.62\sqrt{2.78 \times 10^{-7}}} = 0.271.$$

The contact area between the riser and the casting (i.e., the bottom of the riser) is assumed to be an adiabatic surface; i.e., there is no heat flow by conduction from the riser to the casting. Thus, $z_2 = 0$.

The parameters K_1 and K_2 are determined next.

$$K_1 = \frac{(0.0125)}{(1-0.04)} = 0.0130 \text{ m.}$$

$$K_2 = \frac{(4)(0.04)(3.125 \times 10^{-4})}{\pi(1-0.04)} = 1.658 \times 10^{-5} \text{ m}^3$$

With $\chi = 2$, Eq. (109) becomes

$$D^3 - (0.0130)\left[(4)(0.271)(2) + 0.271 + 0\right]D^2 - 1.658 \times 10^{-5} = 0.$$

Solving $D = 41.4$ mm and $H = 82.8$ mm.

Now we consider casting-riser designs in ceramic shell molds. If we start by selecting a riser that freezes an instant after the casting, then

$$\frac{V_{Rf}}{C_R A_R} = \frac{(V/A)_c}{C_c}. \tag{110}$$

We can follow the steps leading to Eq. (109), but now the asa factors are defined as $e = C_s/C_c$ for the side, $z_1 = C_t/C_c$ for the top, and $z_2 = C_b/C_c$ for the bottom of the riser.

Radiation and convection heat losses from the tops of open risers should also be considered in their design. The heat loss from the top of an open riser in solidification time t is

$$\dot{Q}_t = \left(h + h_r\right)A_t\left(T_M - T_\infty\right)t_s \tag{111}$$

where the subscript t represents the top of the riser. The asa factor is the ratio of the heat loss to that through an equivalent area of the casting surface. For a flat casting surface, the heat loss can be deduced from Eqs. (95) and (96); it is

$$\dot{Q}_c = \frac{kA}{L}\left[\frac{T_M - T_\infty}{\rho' H_f'}\right]\left[\frac{Bi}{1 + Bi}\right]t. \tag{112}$$

Then by comparing Eqs. (111) and (112), we get

$$z_1 = Bi_t(1 + Bi)\rho' H_f'/Bi \tag{113}$$

where Bi_t is the Biot number associated with the heat loss from the top of the riser, defined as

$$Bi_t \equiv \left(h + h_r\right)L/k.$$

Example 21

Derive the asa factor for an open riser attached to a bar-like casting, produced in a ceramic shell mold.

63

Compare Eqs. (97) and (111) and solve for their ratio. Then

$$z_1 = \frac{(h + h_r)A_t(T_M - T_\infty)t_s}{h_c A_c(T_M - T_\infty)t_s}\left[\frac{R_1}{R_2} + \frac{h_c R_1}{k}\ln\left(\frac{R_2}{R_1}\right)\right]$$

With $A_t = A_c$, this simplifies to

$$z_1 = \frac{(h + h_r)}{h_c}\left[\frac{R_1}{R_2} + \frac{h_c R_1}{k}\ln\left(\frac{R_2}{R_1}\right)\right] \qquad (114)$$

Example 22

Refer to *Example 8* and calculate the size of an open end riser, which is attached to an iron casting that is bar-like with dimensions 40 × 40 × 160 (mm). The ceramic shell mold is 10 mm in thickness, and it has thermal properties similar to mullite in Table 10. Assume that $h_c = 58$ W m^{-2} K^{-1} and $f = 0.05$.

From *Example 8*, $h = 12.6$ and $h_r = 106.4$ W m^{-2} K^{-1}.

$$\left[\frac{V}{A}\right]_c = 10 \text{ mm} = 0.010 \text{ m (neglecting the end)}$$

Both the riser and the casting are bar-like and encased in the same shell material; therefore $e = 1$. Also, $z_2 = 1$ because an end riser is specified. The asa factor for the top of the open riser is given by Eq. (114).

$$z_1 = \frac{(12.6 + 106.4)}{58}\left[\frac{40}{50} + \frac{(58)(0.040)}{0.38}\ln\left(\frac{50}{40}\right)\right]$$

$$z_1 = 4.44$$

This extremely high value of the asa factor clearly indicates that alloys or metals with high melting points should not be cast with open risers.

In order to use Eq. (109), we calculate K_1 and K_2.

$$K_1 = \frac{10}{1-0.05} = 10.53 \text{ mm}$$

$$K_2 = \frac{(4)(0.05)(2.56 \times 10^5)}{(\pi)(0.95)} = 1.716 \times 10^4 \text{ mm}^3$$

64

Then (assume $\chi = 2$)

$$D^3 - 10.53 (4 \times 1 \times 2 + 1 + 4.44) D^2 - 1.716 \times 10^4 = 0$$

and

$$D = 142 \text{ mm} \qquad \text{and} \qquad H = 280 \text{ mm}$$

Example 23

   Repeat *Example 22* but assume that the top of the riser is perfectly insulated (i.e., $z_2 = 0$).

   Now the cubic equation for the diameter is

$$D^3 - 10.53 (4 \times 1 \times 2 + 1 + 0) D^2 - 1.716 \times 10^4 = 0$$

then

$$D = 97 \text{ mm} \qquad \text{and} \qquad H = 194 \text{ mm}$$

The yield of this riser is vastly improved over the open top riser. The yield can be improved more by insulating the side of the riser.

B. *Feeding Distance and Riser Location*

   Alloys that solidify in molds of insulating materials freeze over a range of temperatures so that a portion of the casting or the entire casting exists in the solid plus liquid condition for a substantial part of the solidification period. Feed metal, which must compensate for the solidification shrinkage, must flow through a tortuous maze of the dendritic solid. When the resistance to the flow of this metal becomes too great, internal porosity results. This is most evident in alloys with a narrow freezing range that are cast into shapes with uniform cross sections. In these instances, centerline shrinkage may result particularly if a single riser is used to feed an excessive length of the casting. Therefore risers must be located such that the *feeding distance* of the alloy is not exceeded.

   Feeding distance depends on the specific solidification characteristics of the alloy, the mold, and the extent of acceptable porosity in the casting. Feeding distances for some alloys can be found in metal casting textbooks, and metal casters, themselves, develop their own rules based upon experience. Figure 20 shows examples of feeding distances in steel plates and bars made in silica sand molds. Provided that the feeding distance is not exceeded and the risers are properly designed, sound castings result.

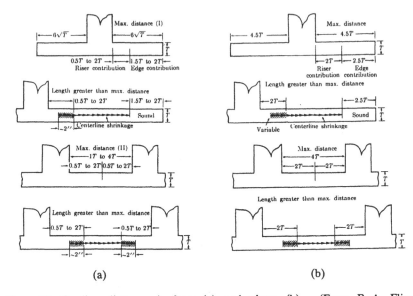

Figure 20: Feeding distances in bars (a) and plates (b). (From R. A. Flinn, *Fundamentals of Metal Casting*, Addison Wesley, Reading, MA, 1963, pp. 54-55.)

In addition to feeding distance, the placement of risers depends on the *freezing order* in the casting. Castings are seldom simple elemental shapes as discussed so far, but complex castings can be subdivided into simple elements, which joined together make the casting. Then the freezing order of the elements must be determined and the riser attached to the element that freezes last, with no element isolated from the riser by an earlier freezing element. The best way to illustrate the placement of risers is with some examples.

Example 24

An aluminum investment casting comprises two bar-like sections connected by a plate. Dimensions are in mm. Design (a) riser(s) for the casting.

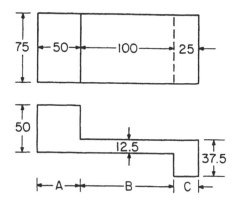

The casting is divided into three elements: A, B and C. Then we estimate the freezing order. For element A:

$$V_A = 1.875 \times 10^5 \text{ mm}^3 = 1.875 \times 10^{-4} \text{ m}^3$$

$$C_A \cong 2.04 \times 10^{-5} \text{ m s}^{-1} \text{ (from } Example \text{ } 17)$$

$$(V/A)_A = \frac{1.875 \times 10^5}{2(50 \times 75) + 2(50 \times 50) + 2(75 \times 50) - 12.5 \times 75} = 9.84 \text{ mm} = 9.84 \times 10^{-3} \text{ m}$$

For element C:

$$V_C = 7.031 \times 10^4 \text{ mm}^3 = 7.031 \times 10^{-4} \text{ m}^3$$

$$C_C = 2.04 \times 10^{-5} \text{ m s}^{-1}$$

$$(V/A)_C = \frac{7.031 \times 10^4}{2(25 \times 75) + 2(37.5 \times 25) + 2(37.5 \times 75) - 12.5 \times 75} = 6.82 \text{ mm} = 6.82 \times 10^{-3} \text{m}$$

For element B:

$$V_B = 9.375 \times 10^4 \text{ mm}^3 = 9.375 \times 10^{-5} \text{ m}^3$$

$$C_B \simeq 1.86 \times 10^{-5} \text{ m s}^{-1}$$

$$(V/A)_B = 6.25 \text{ mm} = 6.25 \times 10^{-3} \text{ m (semithickness)}$$

67

Then the freezing order is

$$\frac{(V/A)_A}{C_A} \quad > \quad \frac{(V/A)_C}{C_C} \quad > \quad \frac{(V/A)_B}{C_B}$$

$$4.822 \times 10^5 \quad > \quad 3.342 \times 10^5 \quad > \quad 3.360 \times 10^5$$

With this simple model our ability to distinguish between the freezing times of elements C and B is not as precise as the numbers indicate. To be safe, we assume it is in the order shown and attach risers to elements A and C.

The riser attached to A must be designed to solidify after a portion of the casting with the modulus of element A, but to feed a portion that has the volume of element A and more than one-half of the volume of element B. So in the riser design equation we use the following values:

$$\left[\frac{V}{A} \right]_C = \left[\frac{V}{A} \right]_A = 9.836 \text{ mm}$$

$$V_c = V_A + \left[\frac{4.822 \times 10^5}{4.822 \times 10^5 + 3.360 \times 10^5} \right] V_B$$

$$= 2.428 \times 10^5 \text{ mm}^3$$

(The ratio in the parenthesis is the proportion of element B that is fed by this riser.)

We assume that the top of the riser is perfectly insulated ($z_1 = 0$) and that a side riser is employed ($e = z_2 = 1$). Also

$$K_1 = \frac{9.836}{(1 - 0.05)} = 10.354 \text{ mm}$$

$$K_2 = \frac{(4)(0.05)(2.428 \times 10^5)}{\pi(1 - 0.05)} = 1.627 \times 10^4 \text{ mm}^3$$

Then, Eq. (107) gives (assume $H = D$):

$$D^3 - 10.354(4 \times 1 \times 1 + 1 + 0)D^2 - 1.627 \times 10^4 = 0$$

Solving: $D = H = 56.8$ mm

Now for the riser attached to element C:

$$\left[\frac{V}{A} \right]_C = 6.818 \text{ mm}$$

$$V_C = 7.0312 \times 10^4 + \left[\frac{3.360 \times 10^5}{4.822 \times 10^5 + 3.360 \times 10^5} \right] 9.375 \times 10^4$$

$$= 10.88 \times 10^4 \text{ mm}^3$$

(Notice that the sum of the volumes fed by both risers equals the total volume of the casting.)

Again $z = 0$, $e = 1$ and $z_2 = 1$. Also

$$K_1 = \frac{6.818}{(1 - 0.05)} = 7.177 \text{ mm}$$

$$K_2 = \frac{(4)(0.05)(10.88 \times 10^4)}{\pi(1 - 0.05)} = 7.291 \times 10^3 \text{ mm}^3$$

Then, with $H = D$:

$$D^3 - 7.177(4 \times 1 \times 1 + 1 + 0)D^2 - 7.291 \times 10^3 = 0$$

Solving: $D = H = 40.4 \text{ mm}$

From a riser design point-of-view, the two risers should be adequate. However, the feeding distance in the plate is probably exceeded (see Fig. 20b), and perhaps another riser (or two) would have to be attached to the plate, itself. A better solution would be to increase the thickness of the plate, and then taper it from element C toward element A.

VI. NUMERICAL METHODS

The solutions in the previous chapters apply to situations of simple geometry. Of course, casting geometry is typically complex so other means of estimating the temperature field during solidification must be sought. It was also mentioned in Chapter II that alloys freeze over a range of temperatures, and in Chapter III we saw that thermal properties generally vary with temperature. When these added complexities are to be included in solving solidification problems, numerical methods should be used. One popular and often used method is called the *finite difference approximation* (FDA).

In this chapter two problems, that were previously discussed in IV.C are reviewed, and then calculated results based on FDA are compared to the exact solutions. Here the major intent is to introduce the reader to the method of FDA, so that simple geometry and constant thermal properties are maintained.

A. Conduction in Mold or Core Surrounded by Metal

Let us look again at the example of a series of alternate platelike sections of solidifying metal and mold and look at the temperature in the mold versus time. We assume that the metal is at its solidification temperature and that the thermal properties of the mold are constant. This situation is depicted in Fig. 17.

Once again the governing differential equation is given by Eq. (10), which is

$$\frac{\partial T}{\partial t} = \alpha \frac{\partial^2 T}{\partial x^2} \tag{10}$$

The boundary conditions are

$$\frac{\partial T}{\partial x}(L,t) = 0 \qquad t \geq 0, \tag{115a}$$

and

$$T(0,t) = T_M \qquad t \geq 0 \tag{115b}$$

The initial condition is

$$T(x,0) = T_i. \tag{115c}$$

In Eqs. (115a,b,c), $x = L$ is at the centerline of the mold, L is the semithickness of the mold, T_M is the solidification temperature of the metal, and T_i is the initial temperature of the mold.* The domain $0 \leq x \leq L$ is subdivided into small segments of length Δx, as shown in Fig. 21. Nodes are numbered consecutively from left to right from 0 to N.

---

*In IV.C, $x = 0$ was at the centerline. The two different coordinate systems can be reconciled with a simple coordinate transformation.

Heat Transfer Fundamentals for Metal Casting
Second Edition with SI Units
Edited by D.R. Poirier and E.J. Poirier
The Minerals, Metals & Materials Society, 1994

70

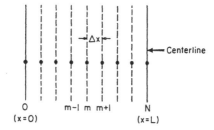

Figure 21: The domain $0 \leq x \leq L$ subdivided into segments.

At a particular time, the gradient at node m and between m and $m+1$ is

$$\left[\frac{\partial T}{\partial x} \right]_{m+\frac{1}{2}} \simeq \frac{T_{m+1} - T_m}{\Delta x}.$$

Similarly,

$$\left[\frac{\partial T}{\partial x} \right]_{m-\frac{1}{2}} \simeq \frac{T_m - T_{m-1}}{\Delta x}$$

For the conduction equation, the second derivative is required; this is

$$\frac{\partial}{\partial x} \left[\frac{\partial T}{\partial x} \right]_m \simeq \frac{1}{\Delta x} \left[\left[\frac{\partial T}{\partial x} \right]_{m+\frac{1}{2}} - \left[\frac{\partial T}{\partial x} \right]_{m-\frac{1}{2}} \right].$$

Therefore,

$$\left[\frac{\partial^2 T}{\partial x^2} \right]_m \simeq \frac{1}{(\Delta x)^2} \left[T_{m-1} - 2T_m + T_{m+1} \right],$$

and Eq. (10) for an interior node becomes

$$\frac{\partial T_m}{\partial t} = \frac{\alpha}{(\Delta x)^2} \left[T_{m-1} - 2T_m + T_{m+1} \right]. \qquad (116)$$

At the centerline $(m = N)$, the finite difference equation is derived in a different manner. The flux to the centerline is

$$-\frac{k}{\Delta x} \left[T_N - T_{N-1} \right],$$

71

and the accumulation of energy is

$$\frac{\Delta x}{2} \rho C_p \frac{\partial T_N}{\partial t},$$

where $\Delta x/2$ is the volume associated with node N. Then at this *insulated surface*, the finite difference equation is written as

$$\frac{\partial T_N}{\partial t} = \frac{2\alpha}{(\Delta x)^2} \left[T_{N-1} - T_N \right]. \qquad (117)$$

For convenience, *normalized variables* are introduced. For time and position, these are exactly the same as the dimensionless numbers that were introduced in Chapter IV. The normalized temperature (u), time (θ) and position (x') are defined as follows:

$$u = \frac{T - T_i}{T_M - T_i},$$

$$\theta = Fo = \frac{\alpha t}{L^2},$$

and

$$x' = \frac{x}{L}.$$

With the normalized coordinates, Eqs. (10) and (115a,b,c) are replaced by

$$\frac{\partial u}{\partial \theta} = \frac{\partial^2 u}{\partial x'^2} \qquad (118)$$

$$\frac{\partial u}{\partial x'} (1,\theta) = 0 \qquad (119a)$$

$$u(0,\theta) = 1 \qquad (119b)$$

and

$$u(x',0) = 0. \qquad (119c)$$

For the interior nodes, the finite difference equation, Eq. (116), becomes

$$\frac{\partial u_M}{\partial \theta} \simeq \frac{1}{(\Delta x')^2} \left[u_{m-1} - 2u_m + u_{m+1} \right], \qquad (120)$$

72

and at the insulated surface ($m = N$) Eq. (117) becomes

$$\frac{\partial u_N}{\partial \theta} \simeq \frac{2}{(\Delta x')^2} \left[u_{N-1} - u_N \right]. \tag{121}$$

For an actual application, the number of nodes should be large, but in order to illustrate the calculation process, we take $N = 4$. Equation (119b) requires $u_0 = 1$; then the difference equations for nodes $1, \ldots N$ are as follows:

$$\frac{\partial u_1}{\partial \theta} \simeq \frac{1}{(\Delta x')^2} \left[u_0 - 2u_1 + u_2 \right] = \frac{1}{(\Delta x')^2} \left[-2u_1 + u_2 \right] \tag{122}$$

$$\frac{\partial u_2}{\partial \theta} \simeq \frac{1}{(\Delta x')^2} \left[u_1 - 2u_2 + u_3 \right] \tag{123}$$

$$\frac{\partial u_3}{\partial \theta} \simeq \frac{1}{(\Delta x')^2} \left[u_2 - 2u_3 + u_4 \right] \tag{124}$$

$$\frac{\partial u_4}{\partial \theta} \simeq \frac{1}{(\Delta x')^2} \left[2u_3 - 2u_4 \right]. \tag{125}$$

The initial temperatures ($\theta = 0$) are $u_1 = u_2 = u_3 = u_4 = 1$.

There are several methods for proceeding; here the so-called *Euler* method is used, in which

$$u_m^{v+1} = u_m^v + \frac{du_m^v}{d\theta} \Delta\theta. \tag{126}$$

It is necessary to add superscripts to indicate time; e.g., u_m^v is the present time and u_m^{v+1} is the future temperature after one time step of duration $\Delta\theta$.

By combining Eqs. (122) through (126) for $m = 1, \ldots N$, with the appropriate superscripts, we get

$$u_1^{v+1} = 1 + (1 - 2P) u_1^v + \qquad P u_2^v$$

$$u_2^{v+1} = \qquad P u_1^v + (1 - 2P) u_2^v + \qquad P u_3^v$$

$$u_3^{v+1} = \qquad P u_2^v + (1 - 2P) u_3^v + \qquad P u_4^v$$

$$u_4^{v+1} = \qquad 2P u_3^v + (1 - 2P) u_4^v \qquad (127a,b,c,d)$$

where $P = \Delta\theta/(\Delta x')^2$. Initially each u_m^v is known ($u_m^0 = 0$), so that Eqs. (127a,b,c,d) comprise four equations with the four unknowns, u_m^{v+1}. After the simultaneous equations are solved, the temperatures at $m = 1, \ldots 4$ are known for time step $v + 1$ ($\theta = \Delta\theta$) and then the process is repeated for as many time steps ($\theta = v\Delta\theta$) as desired.

With the Euler method, numerical oscillations in the solution result if either $\Delta\theta$ or $\Delta x'$ are too large. Specifically, the Euler method requires that $P = \Delta\theta/(\Delta x')^2 \leq 0.5$, in order to prevent numerical oscillations. Table 11 gives calculated temperatures at the insulated surface for $N = 5$ ($\Delta x' = 0.2$) and $N = 10$ ($\Delta x' = 0.1$) with $P = 0.25$. Also shown are the temperatures based on the exact solution. With more nodes the results are closer to the exact solution. With even more nodes, the agreement between the approximated temperatures and the exact temperatures would be even closer. However, with smaller values of $\Delta x'$, $\Delta\theta$ has to be correspondingly reduced in order to meet the criterion of $P \leq 0.5$. Accordingly, more computer time would be required to execute the calculations.

Table 11

Normalized Temperature at the Insulated Surface
with $P = 0.25$

θ	With $\Delta x' = 0.2$	With $\Delta x' = 0.1$	Exact Solution
0.2	0.225	0.229	0.235
0.4	0.522	0.527	0.531
0.6	0.710	0.711	0.714
0.8	0.822	0.824	0.826
1.0	0.892	0.893	0.894

The numerical results can also be used to estimate the amount of energy taken up by the mold (i.e., the domain $0 \leq x \leq L$) as a function of time. There are two methods for carrying out this estimate.

The first method is to track the gradient at the mold surface in contact with the metal. This surface is at $x = 0$ in Fig. 21. For each time step the energy absorbed by the mold can be approximated as the product of the flux and the duration of the time step. By summing these products for all of the time steps, the integrated or total energy absorbed by the mold can be estimated. The accuracy of this method depends on the accuracy of the estimate of the gradient at $x = 0$.

The second method is based upon calculations of the average temperature in the mold. For this particular problem, this method is more accurate than the first because the estimates of temperature are more exact than the estimates of the gradient at $x = 0$.

In the first method, the estimated gradient, in terms of normalized variables, is $(u_1^v - u_0^v)/\Delta x'$. In terms of the original dimensional variables, this is

$$\left[\frac{\Delta u}{\Delta x'}\right]_0^v = \frac{L}{T_M - T_i}\frac{T_1^v - T_0^v}{\Delta x}.$$

Therefore, the flux at the surface is approximated as

$$q_0^v \simeq -k\frac{T_1^v - T_0^v}{\Delta x} = -k\left[\frac{T_M - T_i}{L}\right]\left[\frac{\Delta u}{\Delta x'}\right]_0^v, \tag{128}$$

and the energy absorbed by the mold of area A for one time step is $q_0^v A\Delta t$. Therefore,

$$\hat{Q}_m = \left| A\sum_1^v q_0^v \Delta t \right|$$

$$= Ak\left[\frac{T_M - T_i}{L}\right]\frac{L^2}{\alpha}\sum_1^v \left|\left[\frac{\Delta u}{\Delta x'}\right]_0^v\right|\Delta\theta$$

$$= AL\rho C_p(T_M - T_i)\sum_1^v \left|\left[\frac{\Delta u}{\Delta x'}\right]_0^v\right|\Delta\theta \tag{129}$$

As before the energy given up by the solidified portion of the casting (assumed to be pure metal) is

$$\hat{Q}_C = V\rho'H_f', \tag{130}$$

where V is the volume solidified next to the casting-mold surface. With $\hat{Q}_c = \hat{Q}_m$, Eqs. (129) and (130) finally give

$$\frac{(V/A)}{L\gamma} = \sum_1^v \left|\left[\frac{\Delta u}{\Delta x'}\right]_0^v\right|\Delta\theta \tag{131}$$

where γ is defined by Eq. (84). Equation (131) is the approximate counterpart to the exact solution given by Eq. (90).

In the second method, the average normalized temperature at a given normalized time is calculated. The volume of material associated with the nodes $m = 0$ and $m = N$ is $0.5\,\Delta x'$, so the average normalized temperature is

$$\bar{u}^{\nu} = \frac{1}{N}\left[\frac{1}{2}\left(u_0^{\nu} + u_N^{\nu}\right) + \sum_{m=1}^{N-1} u_m^{\nu}\right] \tag{132}$$

The average dimensional temperature is

$$\bar{T}^{\nu} = \bar{u}^{\nu}\left(T_M - T_i\right) + T_i,$$

so that the energy absorbed by one-half of the mold must be

$$\hat{Q}_m = AL\rho C_p\left(\bar{T}^{\nu} - T_i\right)$$

$$\hat{Q}_m = AL\rho C_p\left(T_M - T_i\right)\bar{u}^{\nu} \tag{133}$$

with $\hat{Q}_m = \hat{Q}_c$, we combine Eqs. (130) and (133); the result is

$$\frac{(V/A)}{L\gamma} = \bar{u}^{\nu}. \tag{134}$$

Of course, Eq. (134) is another approximate counterpart to the exact solution given by Eq. (90).

Approximations of $V/AL\gamma$ based on Eqs. (129) and (134), with $P = 0.25$ and $N = 10$, are given in Table 12. Also shown are values of $V/AL\gamma$ calculated from the exact solution, Eq. (90). The table shows that the estimates based on calculating the average temperature are more accurate than are those based on estimating the gradient at the mold-casting interface.

Table 12

Amount of Solidification as $V/AL\gamma$ versus θ
with $P = 0.25$ and $N = 10$

θ	Finite Difference Approx.		Exact Solution
	Therm. Grad.	Avg. Temp.	
0.04	0.163	0.231	0.229
0.2	0.434	0.506	0.505
0.4	0.626	0.699	0.698
0.6	0.742	0.817	0.817
0.8	0.814	0.888	0.886
1.0	0.857	0.932	0.930

B. *Conduction in Shell Mold with Heat Loss at the Surface*

The shell mold has a thickness L. In this case, the surface in contact with the metal is located at $x = 0$, and the outer surface of the mold is at $x = L$ (see Fig. 22). Also, the normalized temperature is selected so that $u = 1$ and $u = 0$ correspond to $T = T_M$ and $T = T_\infty$, respectively; thus

$$u = \frac{T - T_\infty}{T_M - T_\infty}.$$

By selecting $\theta = \alpha t / L^2$ and $x' = x/L$, then the heat conduction equation is Eq. (118). The boundary conditions and initial condition are as follows:

$$u(0,\theta) = 1,$$

$$\frac{\partial u}{\partial x'}(1,\theta) + Bi\, u(1,\theta) = 0,$$

$$u(x',0) = \frac{T_i - T_\infty}{T_M - T_\infty} = u_i. \qquad (135\text{a,b,c})$$

The temperature at nodes 1 through N-1 are approximated by Eq. (120). By writing an energy balance for node N (with a volume of $\Delta x'/2$), we get

$$\frac{\partial u_N}{\partial \theta} = \frac{1}{(\Delta x')^2}\left[-2(1 + Bi\,\Delta x')u_N + 2u_{N-1} \right] \qquad (136)$$

with $Bi = hL/k$.

In order to advance the solution in time, we again utilize the Euler method and combine Eq. (126) with Eqs. (120) and (136). For nodes 1 through N-1, we get

$$u_m^{\nu+1} = Pu_{m-1}^\nu + (1 - 2P)\,u_m^\nu + P\,u_{m+1}^\nu, \qquad (137)$$

and for node N

$$u_N^{\nu+1} = 2P\,u_{N-1}^\nu + \left[1 - 2P(1 + Bi\,\Delta x')\right] u_N^\nu. \qquad (138)$$

As before $P = \Delta\theta/(\Delta x')^2$, and the superscripts ν and $\nu+1$ represent the present and next time, respectively.

By selecting $P = 0.25$ and $\Delta x' = 0.1$, then $\Delta\theta = 0.0025$; with $u_0 = 1$ the set of equations becomes:

$$u_1^{\nu+1} = 0.25 + 0.5\ u_1^\nu + 0.25\,u_2^\nu \qquad\qquad ; m = 1$$

$$u_m^{\nu+1} = \qquad 0.25\,u_{m-1}^\nu + 0.5\ u_m^\nu + 0.25\,u_m^\nu \qquad ; m = 2, 3, \ldots 9$$

$$u_{10}^{\nu+1} = \qquad\qquad\qquad 0.5\ u_9^\nu + (0.5 - 0.125\,Bi)\,u_{10}^\nu \quad ; m = N = 10$$

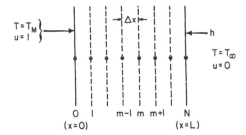

Figure 22: Shell mold of thickness L subdivided into segments for estimating temperature during heat conduction.

Some calculated temperature distributions are shown in Fig. 23 for two initial values of u_m^0 and two Biot numbers. Notice that steady state is achieved in a relatively short period of $\theta = 1$, which corresponds to a real time of only a few or several minutes. Notice, too, that the temperature of the exterior surface of the mold decreases and then increases to its steady state value in some circumstances (e.g., Fig. 23a).

Again the thermal gradients at $x' = 0$ derived from the temperature distributions can be used to calculate the total energy extracted from the casting. Hence to estimate the thickness solidified from the mold wall, Eq. (131) applies. These estimates are shown in Fig. 24.

C. *Closing Remarks*

In the two previous sections, a numerical method, called finite difference approximations (FDA), was used to calculate temperature distributions in shell molds of simple geometry. The examples were selected with the intention of introducing the numerical method; hence, the examples were kept simple. The thermal properties of the mold and the heat transfer coefficient at the exterior surface of the mold were constants. In practice, however, these properties depend on temperature.

It should be noted that variable properties can be included in the FDA method by updating the properties with the current temperatures as the calculation proceeds from one time step to the next. The FDA method can also be extended to two dimensional and three dimensional heat conduction problems so that complex casting geometries can be analyzed.

As mentioned in II.B, alloys solidify over a temperature range; in such cases, nodes are assigned in the solidifying casting as well as in the mold. Sophisticated computer codes can be used to show the positions of isotherms in both the mold and casting, to assist the casting engineer in locating "hot spots," designing and locating risers, estimating solidification times, and in predicting microstructural features and physical properties that depend upon the temperature history during solidification.

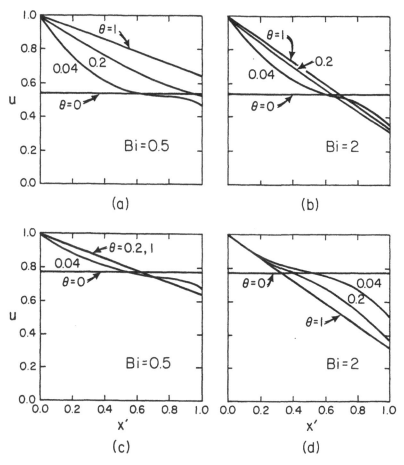

Figure 23: Calculated temperature profiles in a shell mold of finite thickness with heat loss from the surface. (a) Preheat temperature $u_m^0 = 0.536$ and $Bi = 0.5$; (b) $u_m^0 = 0.536$ and $Bi = 2$; (c) $u_m^0 = 0.778$ and $Bi = 0.5$; (d) $u_m^0 = 0.778$ and $Bi = 2$.

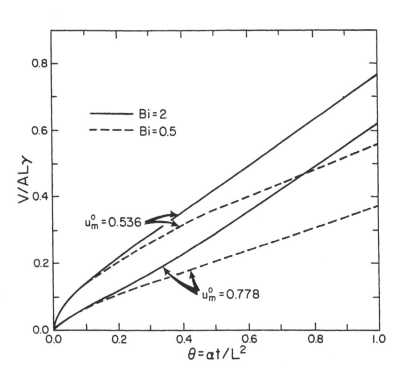

Figure 24: Estimate of thickness solidified (V/A) versus time in shell mold of finite thickness with heat loss from the exterior surface.

SUGGESTED READINGS

J. P. Holman: *Heat Transfer*, 6th ed., McGraw-Hill, New York, NY, 1986.
This is a widely used textbook in many mechanical engineering programs.

F. P. Incropera and D. P. DeWitt: *Fundamentals of Heat and Mass Transfer*, 3rd ed., John Wiley & Sons, New York, NY, 1990.
This, too, is widely used in mechanical engineering programs.

G. H. Geiger and D. R. Poirier: *Transport Phenomena in Metallurgy*, Addison-Wesley, Reading, MA, 1973.
This text comprises fundamentals of fluid dynamics, energy transport and mass transport with applications in metallurgical engineering. One chapter is devoted to solidification heat transfer.

H. S. Carslaw and J. C. Jaeger: *Conduction of Heat in Solids*, 2nd ed., Oxford University Press, Oxford, U.K., 1959.
"Carslaw and Jaeger" is a mathematical treatise of solutions to the conduction heat transfer equation. It is well organized and can also be used a catalog of analytical solutions. It is considered to be a classic by many workers in heat transfer.

G. E. Myers: *Analytical Methods in Conduction Heat Transfer*, McGraw Hill, New York, NY, 1971.
Myers wrote this text in a "self-teaching" style. His chapters on numerical methods are excellent, especially for readers who need a review of matrices.

R. D. Pehlke, A. Jeyarajan and H. Wada: *Summary of Thermal Properties for Casting Alloys and Mold Materials*, National Science Foundation, Grant No. DAR78-26171, Dec. 1982.
This report gives thermal properties as a function of temperature for metals, alloys, and mold materials.

D. R. Poirier and N. V. Ghandi: *AFS Trans.*, vol. 84, 1976, pp. 577-584.
The riser design equation is reviewed and applied in an optimization scheme for minimizing the manufacturing cost associated with risers.

H. Huang, J. T. Berry, X. Z. Zheng and T. S. Piwonka: "Thermal Conductivity of Investment Casting Ceramics," *37th Annual Technical Meeting, Investment Casting Institute*, 1989.
Thermal conductivity data on ceramic shell molds are not generally available. From its title, it is obvious that this is an important reference for investment casting engineers.

H. F. Taylor, M. C. Flemings and J. Wulff: *Foundry Engineering*, John Wiley & Sons, New York, NY, 1959.
Although this book is rather old, the chapter on risering and riser design is still very appropriate.

Heat Transfer Fundamentals for Metal Casting
Second Edition with SI Units
Edited by D.R. Poirier and E.J. Poirier
The Minerals, Metals & Materials Society, 1994

81